Watarrka Waterhole Wildlife

Jenny Davis

Watarrka Waterhole Wildlife by Jenny Davis

Published by Waterpenny Press

NATIONAL LIBRARY OF AUSTRALIA

A catalogue record for this book is available from the National Library of Australia

ISBN: 978-0-6456070-0-0

Contributions
Content - Jenny Davis
Typesetting & layout - James Friend
Editing - Feodora Kojevnikov

This publication can be cited as Davis JA (2022) Watarrka Waterhole Wildlife, Waterpenny Press, Darwin, Australia.

About This Book

This book is first and foremost a book of photographs. It contains the best and most interesting images of wildlife captured on camera traps placed at waterholes in Watarrka National Park, central Australia. Camera traps are cameras that are automatically triggered, so they are a non-invasive way of detecting animals and recording their behaviour. Each camera contains an infra-red sensor that is activated by movement. It is the motion of one or more animals at the waterhole that triggers the camera. In some wildlife research projects a bait is placed near a camera trap to act as an attractant. Baits were not needed here because the Watarrka waterholes provide a source of freshwater that is attractive to species that live in a very arid region. The images have revealed that many species visit the waterholes not just to drink, but also to bathe and to find food. In some cases, they may just be passing through because the waterholes are connected by dry streambeds that can act as pathways for travel across the Park. The bare rocks and sand are easier for reptiles to walk through than the adjacent grasses and shrubs. The streamside vegetation can also act as a dispersal corridor for small and medium-sized birds because it provides protective cover from aerial predators.

This book can be used as a mini field guide because the common, scientific and Indigenous names (where known) of species are provided, together with some accompanying notes. If you want to read more, the references listed at the end of this book provide additional information about the Park and some of the scientific studies that have been undertaken in central Australia.

This book has been a collaboration between myself (a freshwater ecologist working on arid zone waterholes), James Friend who undertook the digital typesetting and layout, and Feodora Kojevnikov who managed the overall production. I am very grateful to James and Dora for the many hours they spent bringing to fruition what I naively thought might be a very simple project but ended up being a much more complex undertaking.

Throughout the preparation of this book we have been mindful of the custodianship of the Traditional Owners, past and present, over many tens of thousands of years, that has supported the resilience and persistence of these waterholes and their wildlife.

Jenny Davis

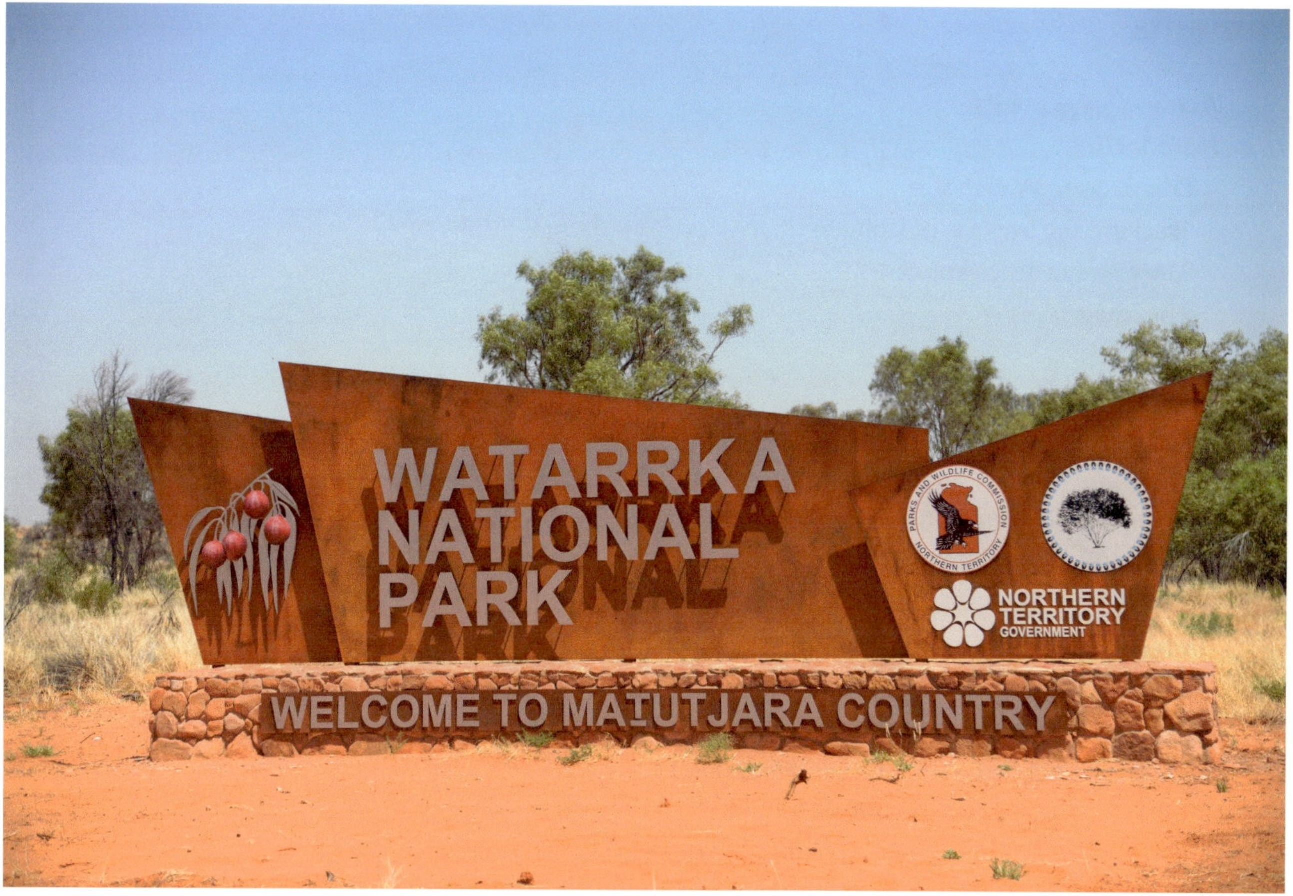
WATARRKA
NATIONAL
PARK
PARKS AND WILDLIFE COMMISSION
NORTHERN TERRITORY
NORTHERN
TERRITORY
GOVERNMENT
WELCOME TO MAIUTJARA COUNTRY

Contents

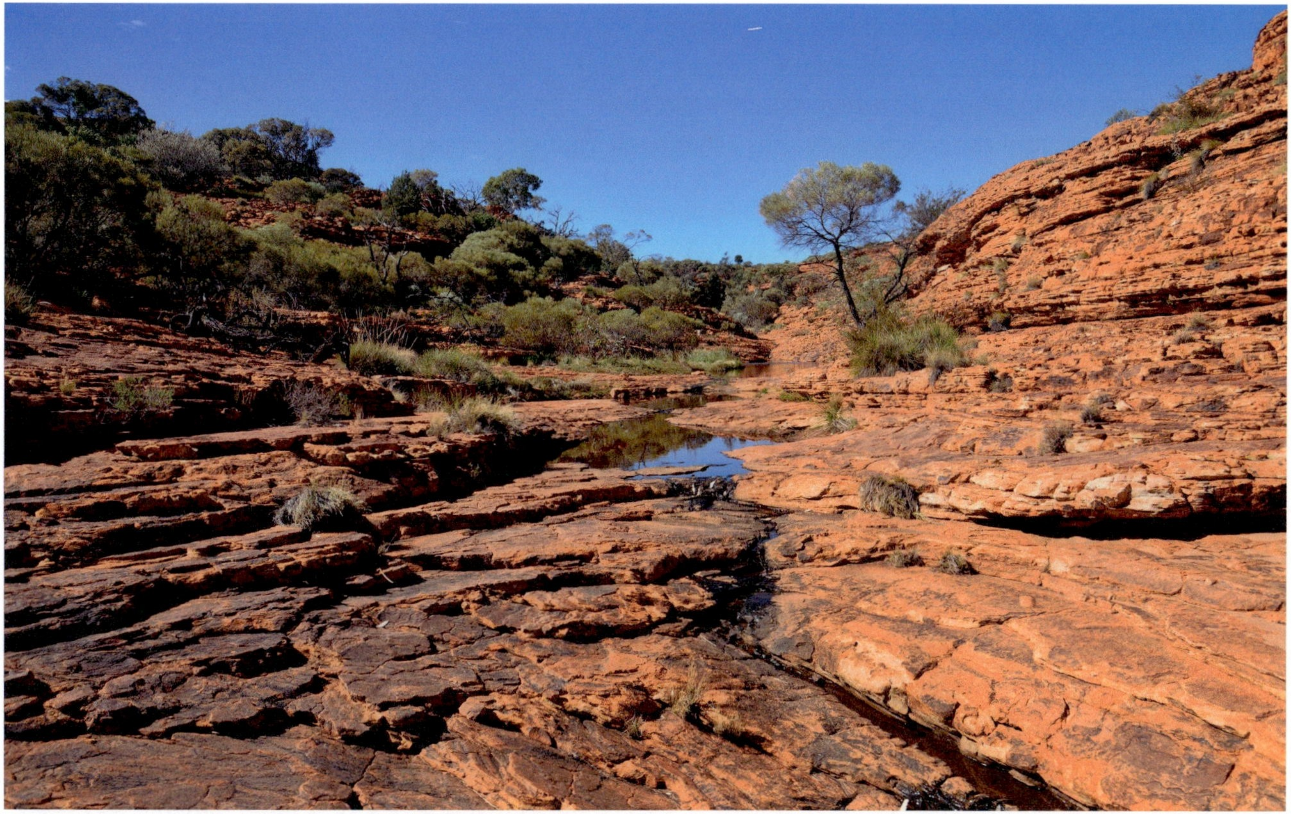

A temporary Watarrka stream flowing after rain.

Acknowledgements

The ongoing support of NT Parks and Wildlife (formerly the Conservation Commission), the Watarrka Traditional Owners, and the rangers who have managed Watarrka National Park since its inception, have all been vital to this project.

Dr Tony Friend provided the initial opportunity to visit the waterholes in 1982 and kindly reviewed the draft manuscript. Many other people who provided invaluable assistance (listed alphabetically by surname) include: Aaron Adamack, Alan Anderson, Pete Beddows, Simon Brown, Rachael Buzio, Jayne Brim Box, Valerie Caron, Stephen Cline, Phil Cowan, Charles Davis, Chris Day, Angus Duguid, Richard Duncan, Fiona Dyer, Glenn Edwards, Elise Furlan, Dave Gibson, Chris Hakanson, Jon Hodgetts, Mirjam Kaestli, Ian Kidd, Peter Latz, Amber Marshall, Dennis Matthews, Glenis McBurnie, Peter McDonald, Robert Mengler, Ash Murphy, Roanne Ramsey, Mike Rawnsley, Emma Razeng, Robert Ridgeway, Derek Roff, Samara Rogan, Gail Rossi, Lauren Schaller, Christine Schlesinger, Alistair Stewart, Paul Sunnucks, Ross Thompson, Chanai Tuckey, Simon Votto and Gary Weir. Vanessa Solano Rivera re-drew the location map from the Watarrka National Park JMP (2010).

Animal ethics approvals were provided by the University of Canberra Animal Ethics Committee (CEAE 15-15) and the Charles Darwin University Animal Ethics Committee (A17018). Permits for scientific research were provided by the Watarrka National Park Joint Management Committee. Finally, this book could not have been completed without the facilities and support provided by the Research Institute for Environment and Livelihoods, College of Engineering, IT & Environment, Charles Darwin University.

Introduction

At a time when the world is facing multiple and major environmental threats, including global warming and biodiversity loss, it is heartening to know that there are still some places where ecosystems, and the plants and animals they support, remain largely intact. Watarrka National Park is one of these places.

When you arrive at the Park, which lies at the western end of the George Gill Range, you are not only entering a place that will impress you with its magnificent rocky outcrops, intriguing sandstone formations and lushly vegetated gorges, you are also stepping back in time. The cycads and ferns in the shaded gorges, and some of the aquatic species within the waterholes, have persisted from a time when central Australia had a very different climate. These species have origins from a far distant past, some reaching back to when Australia was part of Gondwana. The climate changed as Australia separated from the southern polar landmass and moved northwards. Some species with very old lineages have persisted, while newer species have arrived. Ancient origins and more recent cycles of warmer and wetter conditions, alternating with cool and dryer times, have all influenced the plants and animals that occur in the Park today.

Today's species can cope with the extreme climatic variability that characterises modern day central Australia. Extended droughts can occur and no rain may fall in some years. In contrast, rare but large rain events can create flash floods that inundate large areas. Below freezing temperatures occur at night in winter, while some of the hottest day time temperatures on the planet occur in summer. Some Park species may be well-adapted to persist in a warming world, however, many will be vulnerable.

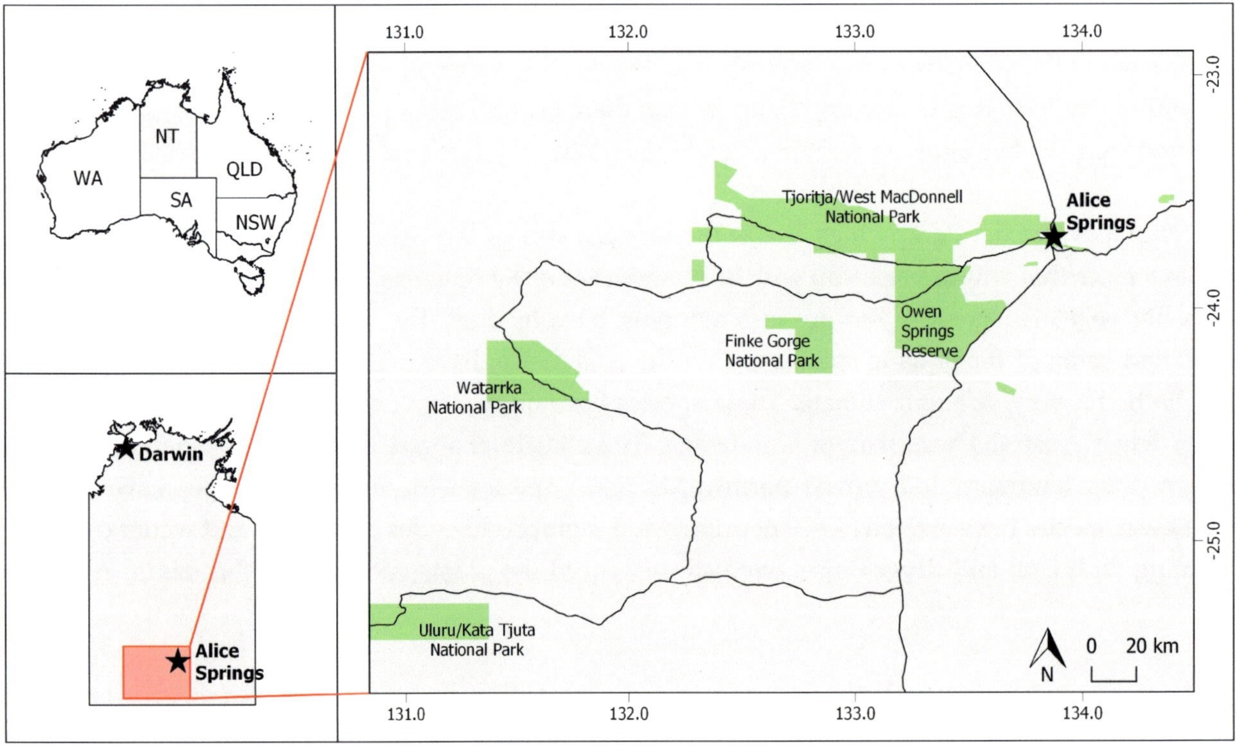

Watarrka National Park is located approximately 300 km southwest of Alice Springs in central Australia. National parks and reserves are indicated in green.

This book provides images of the wildlife captured on camera traps at the Watarrka waterholes. The region has been the home of the Martutjara people for many thousands of years and the Park is now jointly managed by the Traditional Owners and the Parks and Wildlife Commission of the Northern Territory. Watarrka was declared a national park in 1989 and scheduled under legislation as NT parks freehold land in 2005. Joint management plans, published in 2010 and 2018, describe the values of the Park and how its rich cultural, historical, ecological and geological assets will be managed and conserved (Watarrka NP JMP 2010, 2018).

In addition to being a living cultural landscape, this area that contains so much water relative to the arid region it lies within, is immensely important for biodiversity conservation. The shaded gorges, permanent springs, and waterholes provide refuges for many plants and animals. Over 600 species of plants, 20 native mammal species, 126 bird species, 71 reptile species, 7 frog species and 109 invertebrate taxa have been recorded.The spectacular scenery, including Kings Canyon, at the western end of the George Gill Range, also makes the Park a popular tourist destination.

Discovering Waterholes

Gently flowing streams and dripping, moss-covered walls deep within sheltered, cycad-lined gullies, located in an otherwise very arid region, intrigued me on my first visit to the Park in 1983. These freshwater habitats have continued to inspire much of my research ever since. My introduction to the George Gill Range was provided by Peter Latz, a botanist with the Conservation Commission of the NT, with wide-ranging knowledge of the flora, fauna and ecosystems of central Australia.

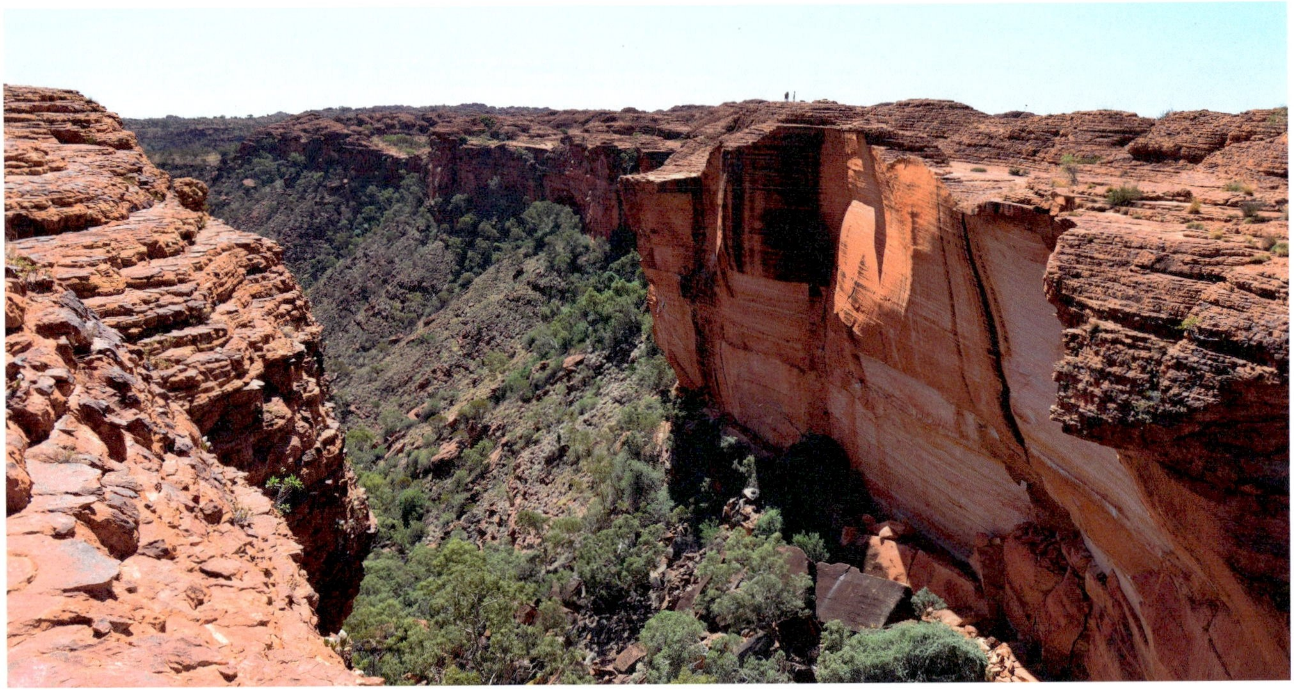

The western wall of Kings Canyon.

Peter, together with other CCNT researchers, Ken Johnson and Mike Gillam, conducted a biological survey of the Kings Canyon area of the George Gill Range in 1981. They considered the Range and its environs to be the most botanically important area in central Australia. They recorded over one third (approximately 158 species) of the central Australian flora. They noted that some of the best stands of desert trees occurred in the area and introduced weed species were largely absent. Cattle were also largely absent and other species, including camels, feral horses and rabbits, were present only in low numbers. They recommended the establishment of a national park (now Watarrka National Park) at the western end of the Range to conserve one of the richest natural areas in the Centre, whilst also supporting the developing tourist industry (Latz *et al.* 1981).

Following Peter's suggestion to visit the Range, I was able to see some of the waterholes on the return leg of a field trip across the NT and WA with Dr Tony Friend and Phillip Fuller. Their project, for the WA Department of Fisheries and Wildlife, was to document the last sightings of the numbat (*Myrmecobius fasciatus*) across inland Australia. Although we only made a very brief visit to the Range, that fleeting glimpse was enough to indicate the unique nature of these arid zone waterholes and to provide the basis for a research proposal. My project proposal, to undertake a survey of the waterholes and their aquatic fauna, was subsequently funded by the Australian Water Resources Advisory Council in 1986. This was the beginning of my research on the Park's aquatic ecosystems that has continued to today.

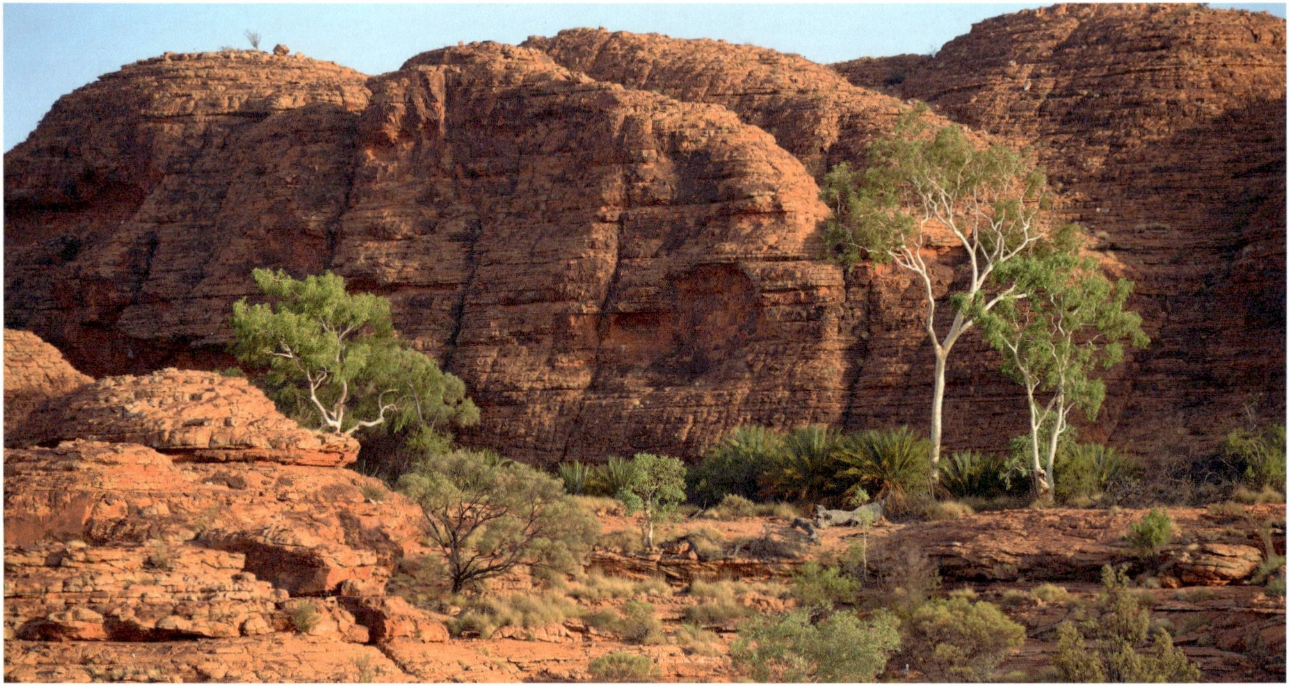

Lush pockets of vegetation dominated by ghost gums (*Eucalyptus papuana*), cycad palms (*Macrozamia rnacdonnellii*), native pines (*Callitris glaucophylla*) and native figs (*Ficus platypoda*) occur where groundwater is present at, or just below the surface in the Park.

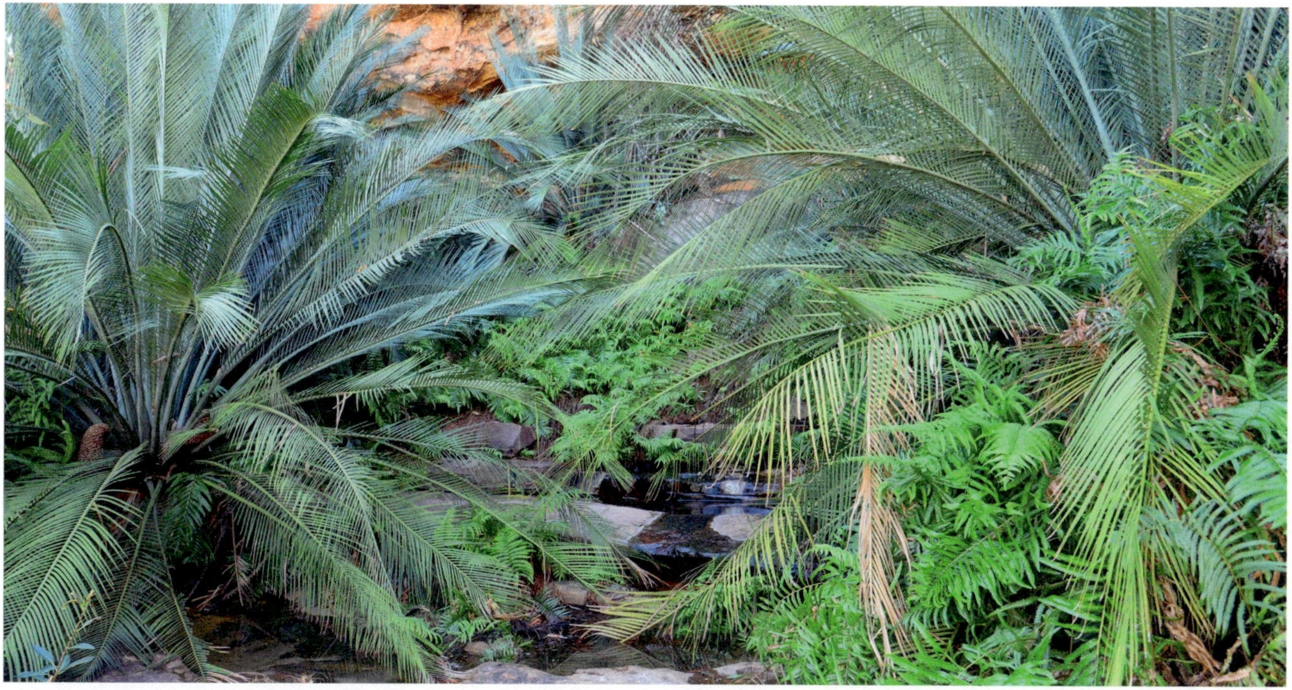

Perennially flowing springs in shaded, moist microhabitats within south-facing gorges, support ferns, mosses and aquatic insects, including water pennies (Psephenidae), mayflies (Ephemeroptera) and caddisflies (Trichoptera). These insects are groups with Gondwanan origins and are closely related to species that are found in cooler, southern and more temperate regions of Australia.

The first written description of the George Gill Range, and its waterholes, was provided in the diary of the explorer Ernest Giles. He named the Range, and some of its features, in 1872, and noted that it was the best watered area within the region. Giles was fortunate to arrive at the Range after substantial rains had fallen (Giles 1889). Twenty-two years later, in 1894, the Horn Expedition visited the George Gill Range as part of the first scientific expedition to central Australia (Spencer 1896). This expedition seems to have arrived at a much drier time. They recorded the presence of aquatic invertebrates but neither fish nor frogs were observed. They collected two new species of mollusc and 'a curious Orthopteran insect resembling a small flattened-out cockroach which adheres almost as closely to the surface of submerged leaves as a limpet to a rock'. This was probably a water penny, the aquatic larva of a small beetle in the family Psephenidae.

Why was this a standout sentence for me from a four volume report on the expedition? It was this group of aquatic insects that I had studied for my PhD research. At that time, in the late 1970's and early 1980's, there were many groups of Australian freshwater invertebrates that were almost unknown, particularly in comparison to the fauna of the rivers and streams of the Northern Hemisphere. My PhD project involved describing adult water penny beetles (Psephenidae) and their aquatic larvae. They occur mainly in the streams and rivers of the Great Australian Divide, from southern Tasmania to northern Queensland (Davis, 1986). The larvae graze the biofilms on rocky stream substrates. They are negatively phototropic (they avoid light) so you will only see them if you turn over the rock or leaf they are sheltering under. The larvae are usually associated with rocky habitats in fast flowing, well-oxygenated streams and rivers, and their presence is considered to be an indicator of good water quality.

The water penny that occurs at the Range, *Sclerocyphon fuscus*, has an aquatic larva and a small adult beetle with limited flying ability. This species is not able to travel across the large tracts of arid land that now separate the Range from the southern regions of Australia where it also occurs. The presence of other stream-dwelling species that are weak fliers, such as mayflies and caddisflies, supports the concept that this is a relictual (surviving from the past) stream fauna. Similar to the cycads, ferns and mosses that occur in the Park, these species of aquatic insects appear to have persisted from a time when central Australia was much wetter than it is today.

Subsequent research at the springs and waterholes in Watarrka NP and Tjoritja/West MacDonnell NP, has shown that freshwater species with Gondwanan origins, first recognised by the Horn Expedition in 1894, occur within central Australia today (Davis *et al.* 1993, Brim-Box *et al.* 2014). This persistence is attributed to multiple factors, including the role of local aquifers (groundwater) in providing permanent freshwater, the remote and inaccessible nature of the sites, and the protection and management provided by their location within national parks.

Tracking the Impacts of Climate Change on Waterhole Wildlife

The Park's springs and waterholes are immensely important because they support much of the region's aquatic and terrestrial biodiversity. Their importance was recognised in 2005 by a comprehensive survey of arid zone NT wetlands and waterbodies undertaken by Angus Duguid and co-workers (Duguid *et al.* 2005). However, although the Park's aquatic biota are largely protected from many of the threatening processes that occur elsewhere, such as habitat destruction, water extraction, pollution, feral animals and invasive plants, the impacts of global warming are all pervasive. Central Australia is warming more rapidly than anywhere else in Australia. Annual maximum temperatures recorded at the Alice Springs meteorological station have increased by 2°C since 1900.

The rainfall regime in central Australia is one of the world's most variable and unpredictable (Murphy *et al.* 2015). Large storms bring flooding rains in either summer or winter, but there can be months, and sometimes years, with little rain. When high summer temperatures combine with prolonged droughts, enormous pressure is placed on the Park's wildlife and the waterholes that support them. This project came about through the need to understand how terrestrial species use the waterholes and how changing climatic conditions may affect them.

The Camera Trap Project

The following images all come from a camera trapping project that started at a Watarrka perennial spring in December, 2013. The project began with advice (and a borrowed camera) from Dr Jayne Brim Box, a scientist with the NT government, who had been using cameras to determine the impacts of camels on a small spring south of Uluru. Jayne was working with traditional owners from the Mutitjulu community who were concerned about the impacts of camels on the spring. It is an important source of drinking water for people travelling between the A̲nangu Pitjantjatjara Yankunytjatjara (APY) lands and Uluru. A camera placed at the spring recorded feral camels arriving at the spring at sunset and drinking the spring dry by dawn. The study provided unequivocal evidence of the damage that camels can do to waterholes, particularly under drought conditions when water is scarce (Brim Box *et al.* 2010). It also indicated the potential for using camera traps to record waterhole use by wildlife in arid inland Australia.

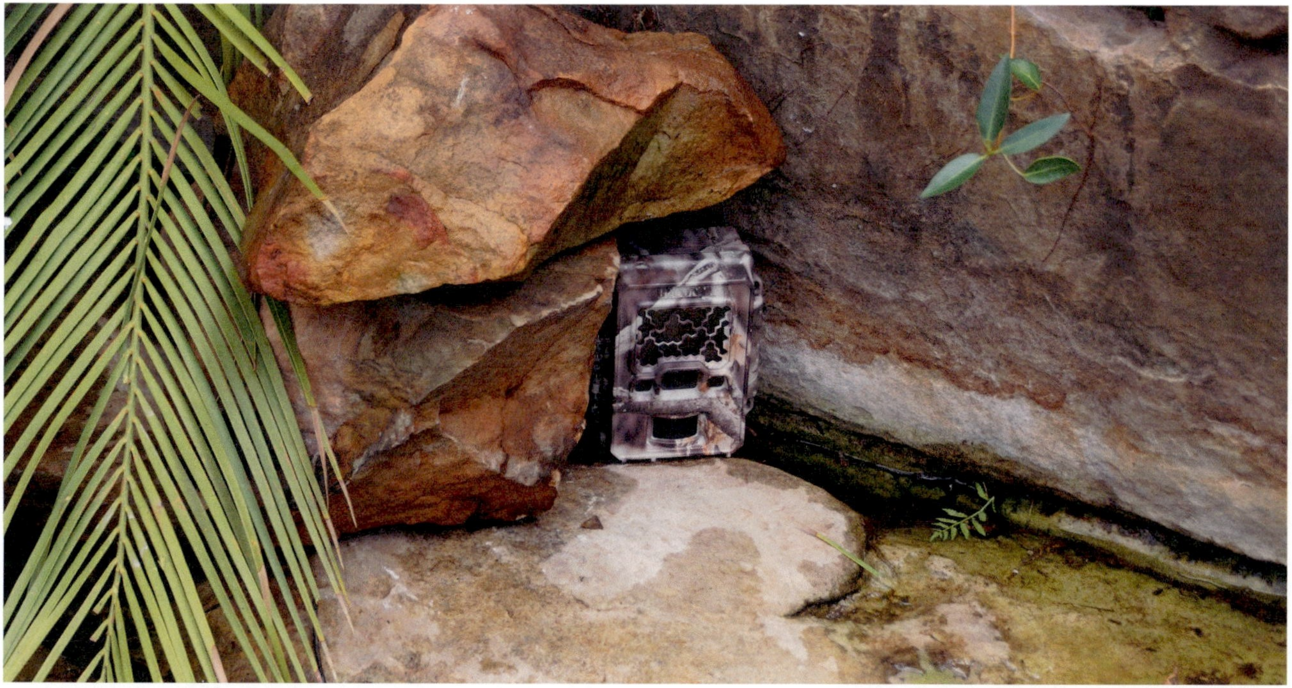

A camera trap was first placed on a rocky ledge on the edge of a Watarrka perennial spring in December 2013.

Passive infrared sensor-triggered Reconyx HyperFire HC600 and HP2X HyperFire 2 cameras have provided the images given overleaf. The cameras are set to operate continuously with three images per trigger event. They are powered by 12 NiMh rechargeable batteries and images are stored on SD cards.

After the successful pilot study, more cameras were added to the perennial spring site. Very occasionally a flash flood will wash a camera downstream. However, the long periods of predictably constant spring flow means that cameras can be placed very close to the water's edge. These provide close up images of the species coming to drink and to bathe.

In addition to the spring site, cameras were installed at a further three waterholes chosen to represent a range of water regimes from permanently wet to almost always dry. These included: a permanent rockhole fed by groundwater seeps and local rainfall; an intermittent stream pool that holds water after rainfall events and in the cooler months when evapotranspiration decreases and the local water table rises; and a temporary stream pool that only holds water for a short time (2-4 weeks) after rain has fallen.

All the camera trap sites are located within the headwaters of Park streams. However, because this is an arid region, where the frequency and timing of rainfall events varies enormously, these streams rarely flow (with the exception of the spring site). These isolated waterholes attract many species (dingoes, kangaroos, wallabies, birds, some reptiles, dragonflies, damselflies, butterflies and wasps) that come to drink, to bathe and, in the case of predators, to find food. They provide ideal sites to record much of the Park's faunal biodiversity and to track changes in wildlife and water availability over time.

The Images

Hundreds of thousands of images have been collected since the first waterhole camera was installed at a Watarrka perennial spring in late 2013. Although very few of the images are good enough to publish, many contain enough information to identify the species present. Over time they have provided an enormous amount of information about the occurrence, condition and behaviour of many of the vertebrate species, and some invertebrates, that occur in the Park. They also provide information on the role of waterholes as ecological refuges during seasonally hot and dry times, and droughts. The very best of the images, from 2014 to 2022, are provided in the following pages.Common names and scientific names provided for all species shown here are based on recently accessed information provided in the Atlas of Living Australia (https://www.ala.org.au). A list of all the species recorded by the cameras, from 2014 to 2022, are provided at the end of the book.Indigenous species names, where known, have been included in the text.

@twitter

Research data can take years, and sometimes decades to be collected and published as papers in scientific journals. Access to these journals is usually only via universities and research institutions, and so papers are often only read by research scientists. However, social media platforms provide the means for rapid communication, albeit in a much more truncated way, to a much larger audience. The best images (from a photographic perspective) and those showing interesting or unusual animal behaviour, have been posted to @waterpenny10 on Twitter.

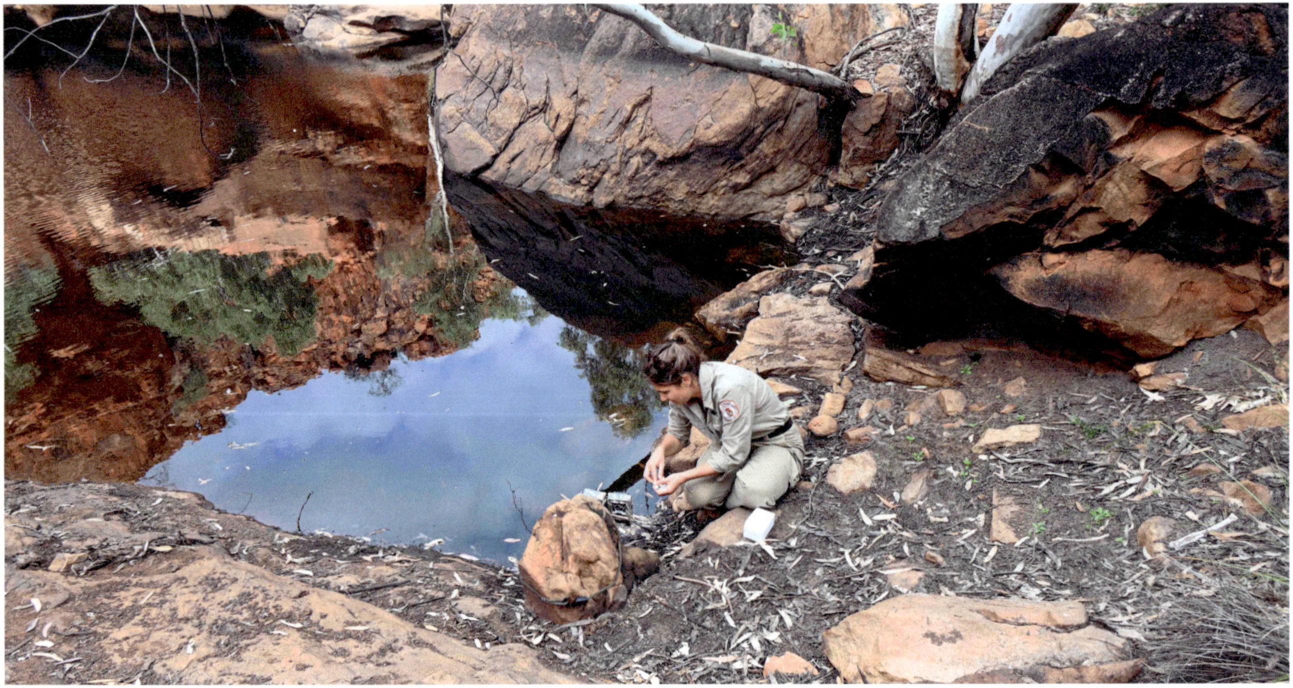

NT Parks and Wildlife ranger Lauren Schaller, changing the batteries in a camera at the edge of the permanent rockhole.

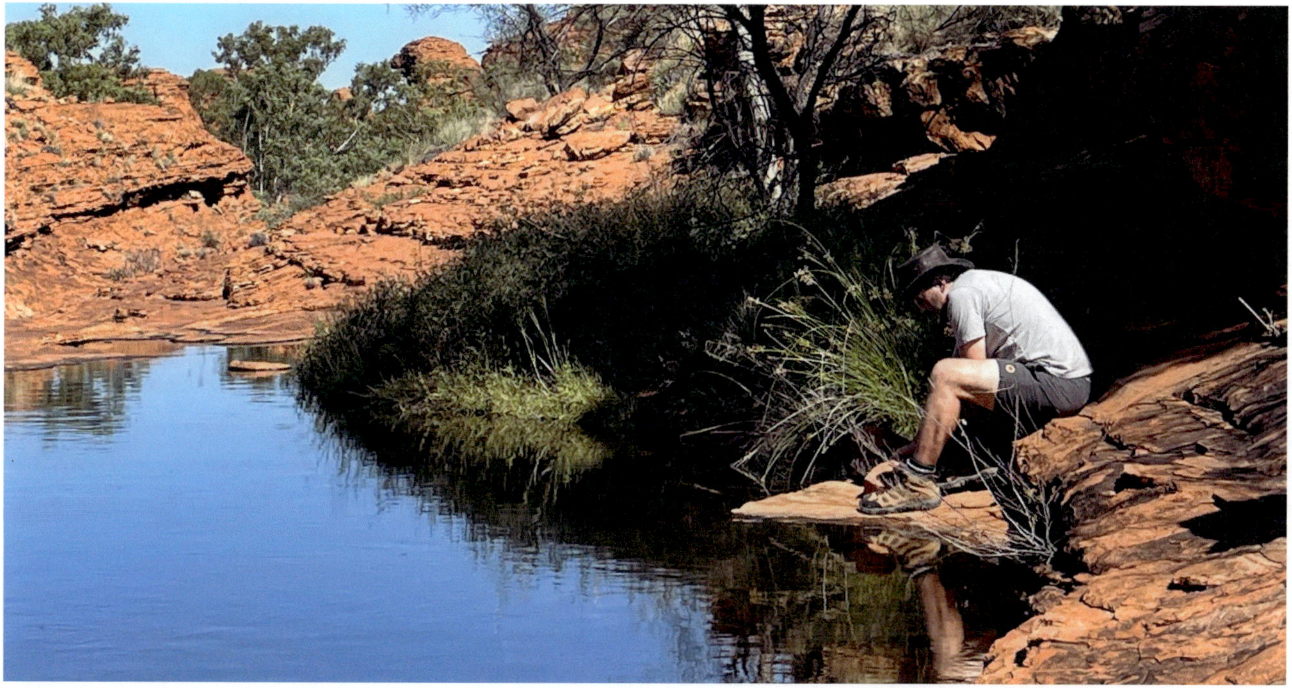

Fieldwork volunteer James Friend, installing a camera at the edge of the recently flooded pool on the temporary stream.

Mammals

Dingo, *Canis familiaris dingo* or *Canis dingo*

Dingoes are Australia's largest land predator and an important keystone species. They can play a major role in maintaining biodiversity through the regulation of terrestrial food chains (Letnic *et al.* 2009). However, although dingoes are recognised as being an iconic part of the Australian mainland fauna (they do not occur in Tasmania) there is much debate about their scientific name (Smith *et al.* 2019). In the past all animals that could interbreed were considered to belong to the same species, this is known as the 'biological' species concept. This places the dingo in the same group as domestic dogs, wolves, coyotes and other canids. However DNA analysis tells a different story. A recent genomics study (Field *et al.* 2022) has found that dingoes are genetically different from both wolves and domestic dog breeds, and can be recognised as a separate species. Where genetic mixing has occurred with domestic dogs, it has been very recent.

The large number of dingo images recorded on the waterhole cameras indicates that the Park potentially supports a healthy population. However, it is not possible to determine the actual size of the dingo population from the design of this camera trapping project. Undoubtedly water is the drawcard, dingoes need to drink and the hot dry summers of the region, together with extended droughts, mean that waterhole visits are frequent. The waterholes are also a source of prey. Small mammals and reptiles that are an important part of the dingo diet are also attracted to the water. The euros that visit the waterholes to drink in the warmer months can also become prey when dingoes hunt as a pack.

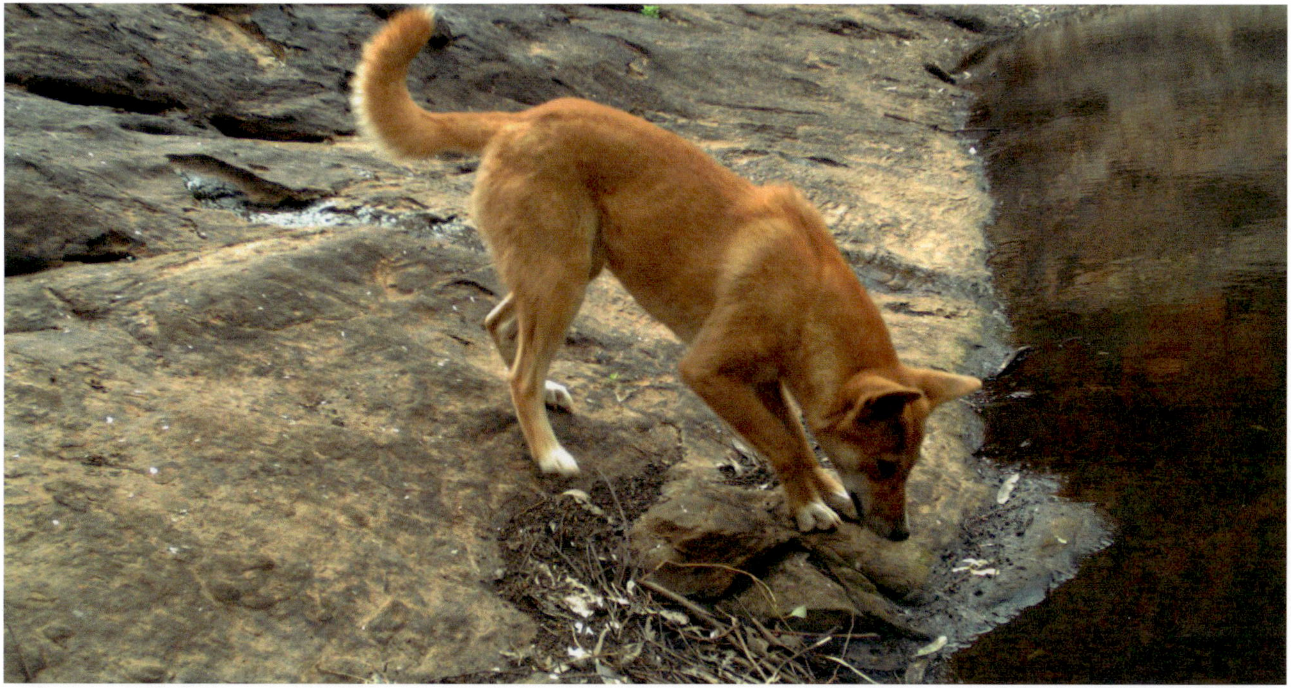

Classic dingo colours: a golden coat and white feet. The healthy appearance of this dingo suggests that the preceding year has been a good one with enough water to support both predators and prey.

Forty camera trap images of dingoes were posted on the Twitter account @waterpenny10 between 2017 and 2020. Our analysis revealed that these images were the most liked and most retweeted of all posts from the Watarrka camera traps. Some tweets triggered a discussion as to whether Watarrka dingoes were wild dogs, i.e. dingo/domestic dog hybrids. The remoteness of the Park suggests that most dingoes can be assigned genetically to *Canis dingo,* although it is most likely that some hybridisation with local dogs has occurred.

Not all Watarrka dingoes are golden, occasionally a black and tan dingo is recorded on a camera. This image sparked a Twitter discussion in 2020 as to whether these are wild dogs, not dingoes. The matter was clarified by @KDNewman87, who noted that the first Western painting of a dingo by George Stubbs, in 1772, titled Portrait of a Large Dog, was of a black dingo from NSW at a time before the arrival on the continent of domestic dogs from Europe.

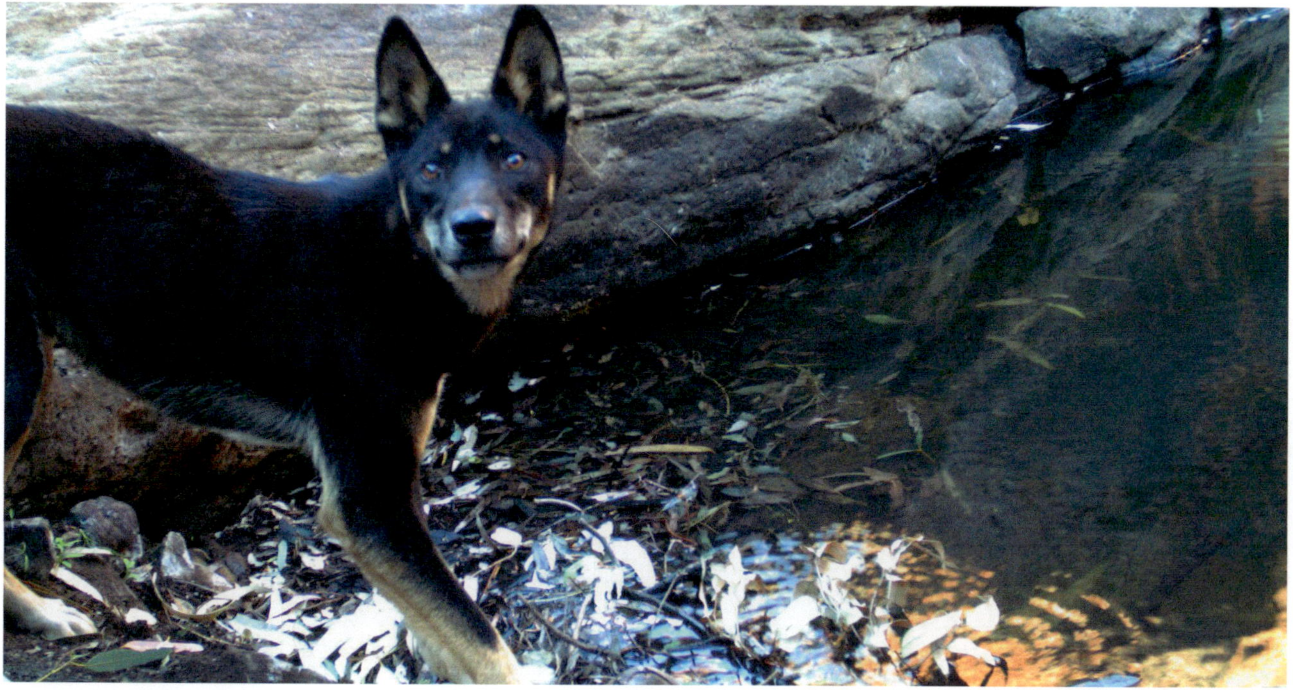

A black and tan dingo.

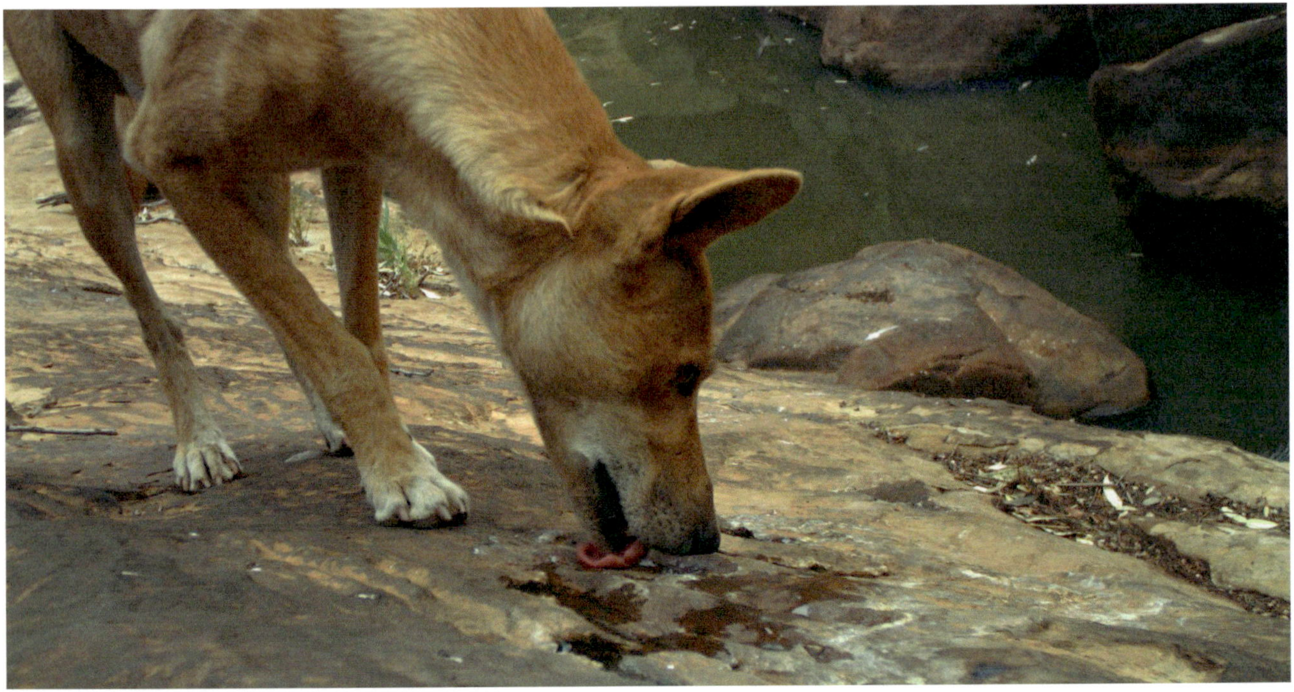

A hot summer, almost no summer rain, and declining water quality in the permanent rockhole, means that groundwater seeping from the adjacent rock wall becomes the favoured water source for many species.

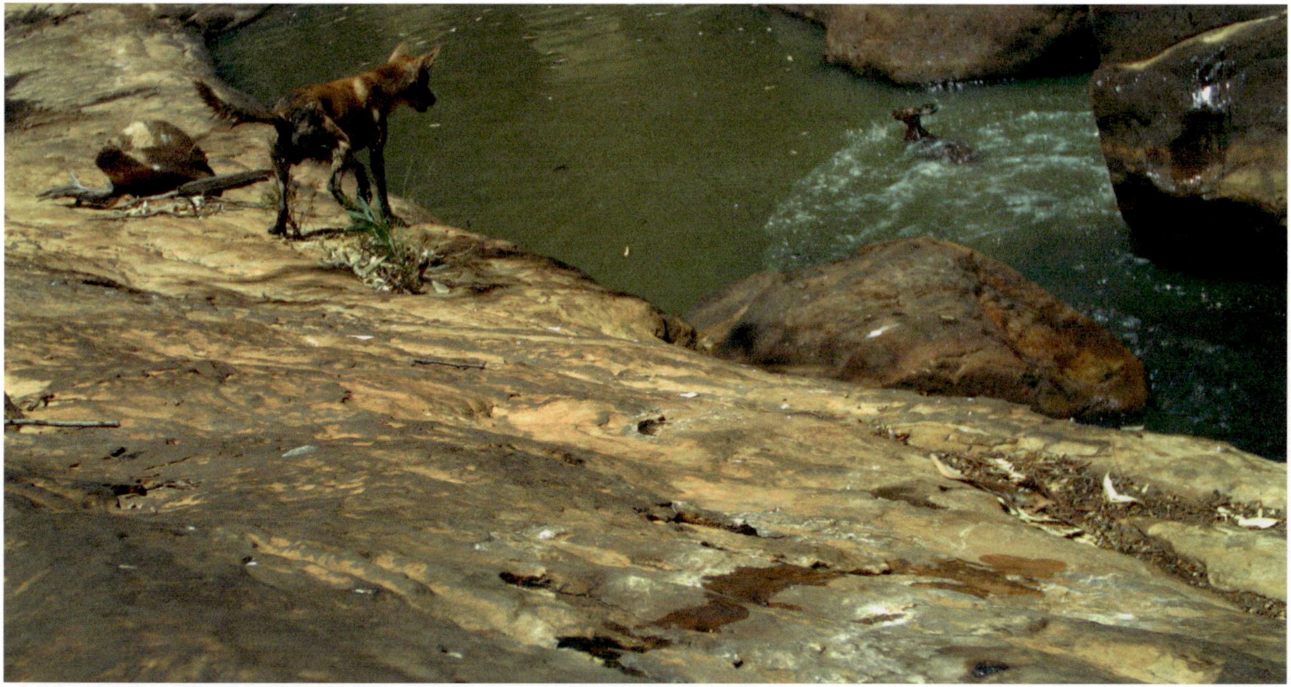

Some of the most confronting images recorded from the Park occurred on the cameras at this permanent rockhole. They document a pack of dingoes chasing a euro which initially escaped by jumping into the water.

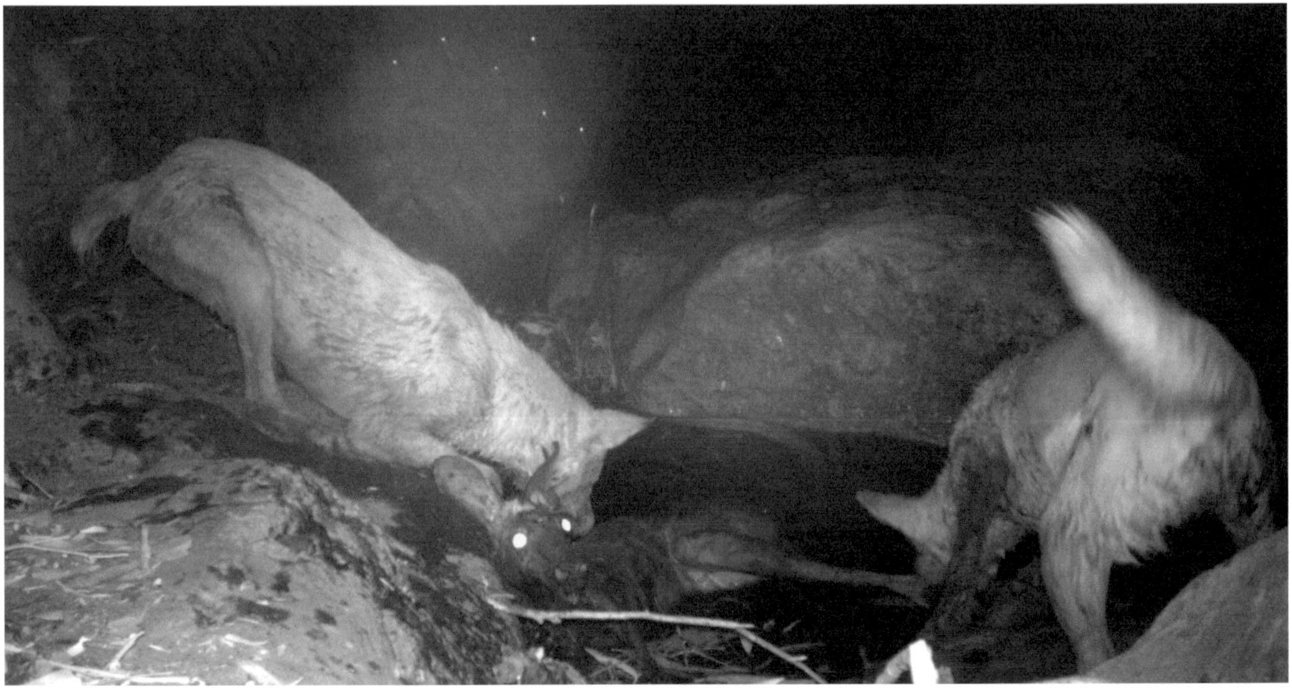

Five days after the images of the dingoes hunting as a pack appeared at the drying rockhole more images revealed that the euro had become their prey.

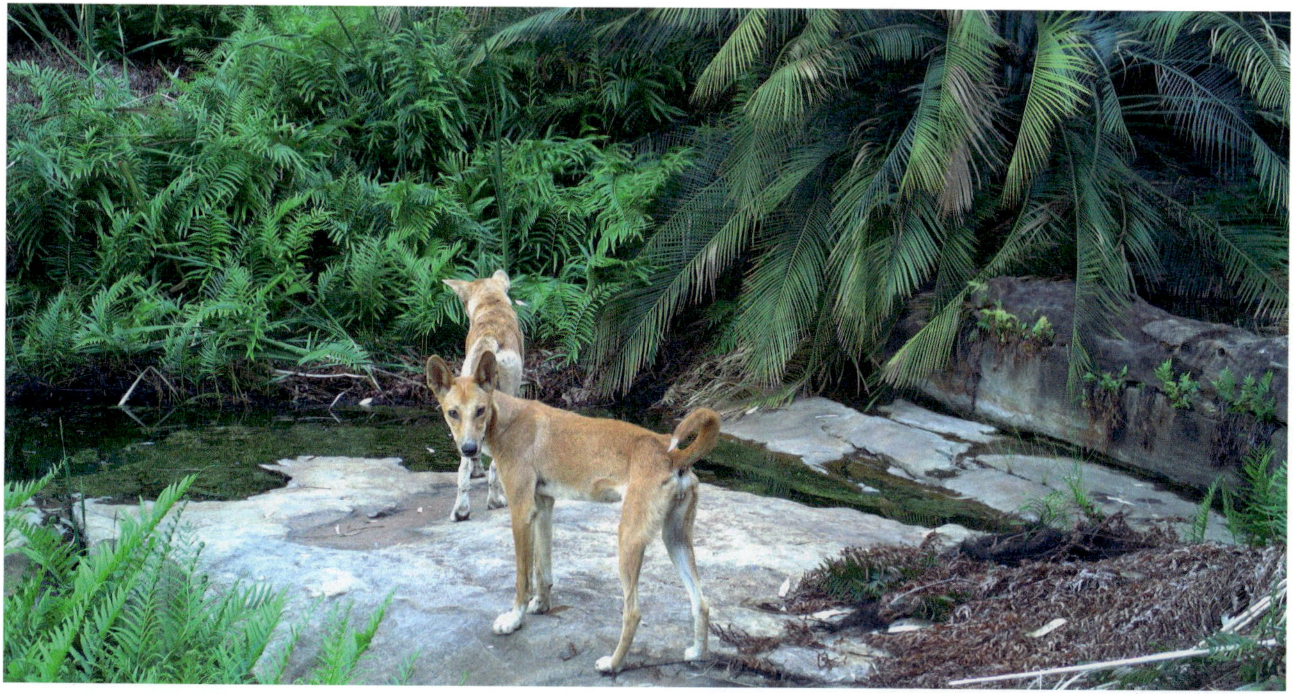

An additional camera in a higher position was installed at the perennial spring in 2020. It provides a wider view of the dingoes and other species that visit the rocky pools amidst the cycads and ferns that thrive in this shaded gully.

Marsupials

Euro/Kulpirrpa, Hill Kangaroo, Common Wallaroo, *Osphranter robustus*

Euros/kulpirrpa, or common wallaroos, tend to occupy rocky country and the waterhole cameras have provided much evidence that a large population of euros lives within the Park. This species is usually nocturnal (active at night) and there are many images of males, females with joeys, and young euros, drinking at the waterholes after darkness has fallen. However, the following images indicate that many are also active during the day. Euros are herbivores and the grasses, herbs and shrubs that occur along the creek lines provide a good source of food, especially after rain has fallen. Euros are usually solitary animals, but female euros are often recorded with a young offspring nearby. Euros need to drink free water, so the waterholes are important for their survival. However, the waterholes are also a place of risk, where euros are vulnerable to attack from their main predators, dingoes and wedge-tailed eagles.

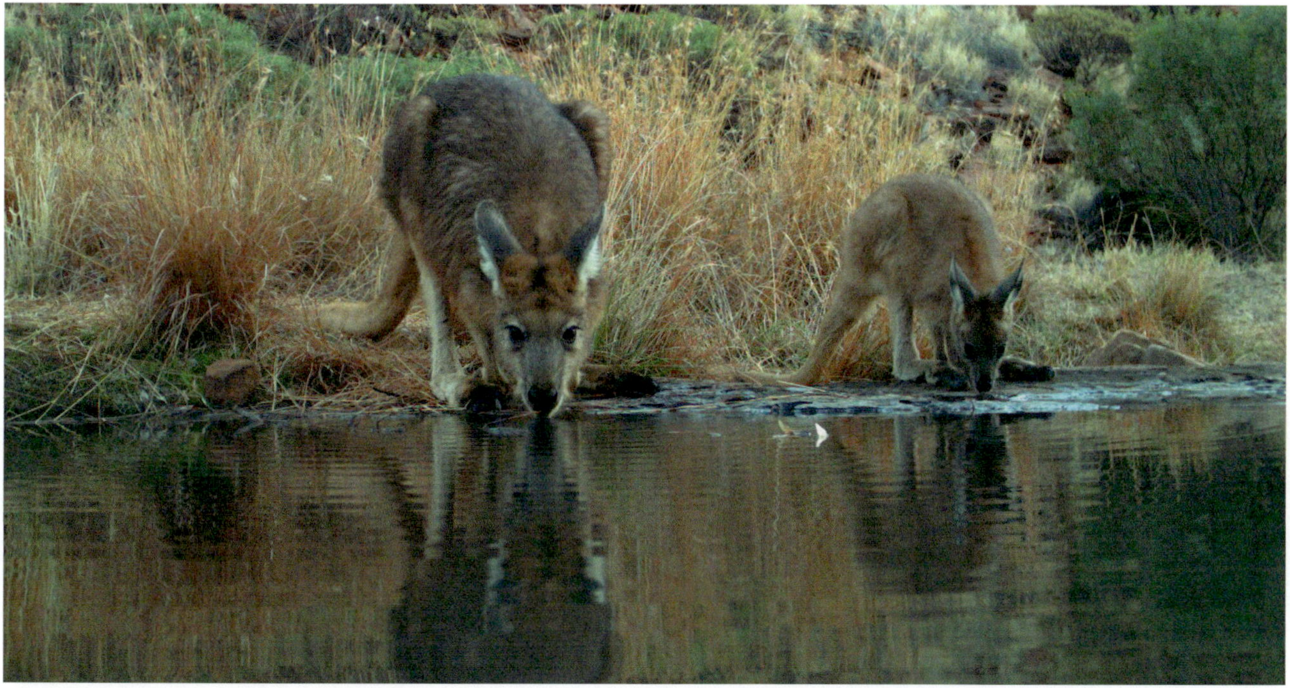

This female euro and her young offspring have come to drink at the shallow intermittent pool in the late afternoon.

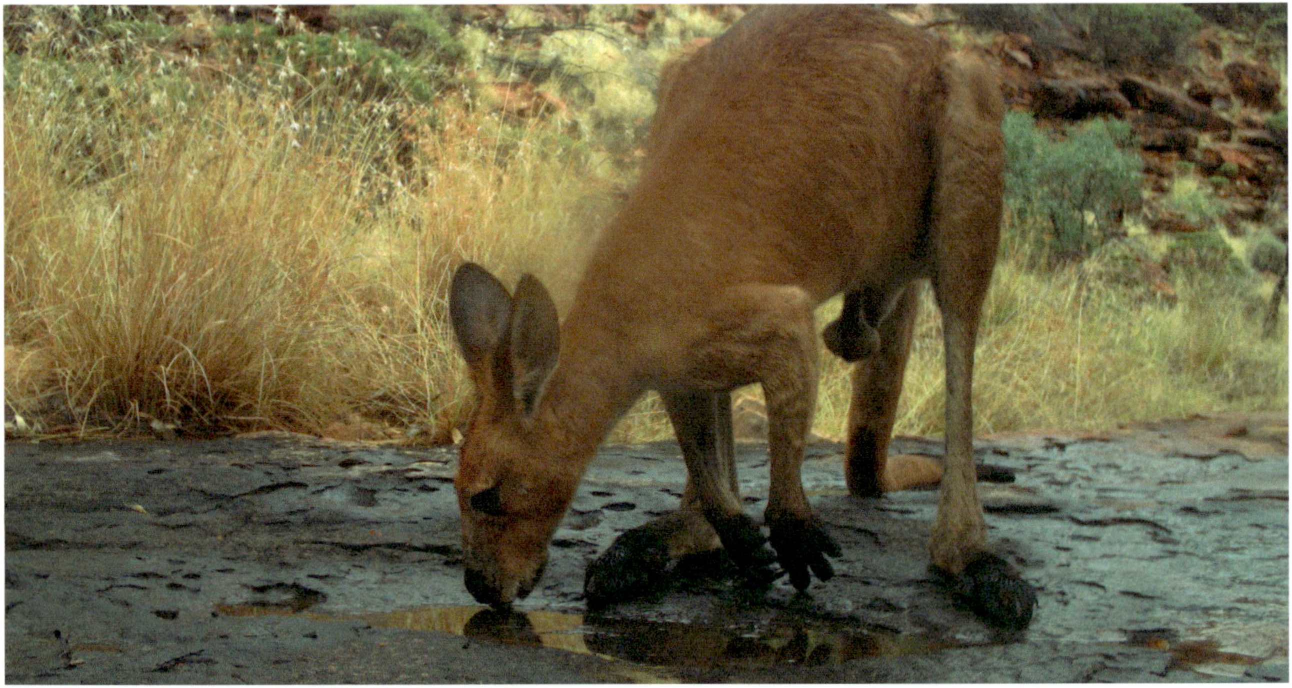

On a hot day a small drying pool provides enough water for this male euro, kulpirrpa, while the local grasses (in the background) provide an abundant food source.

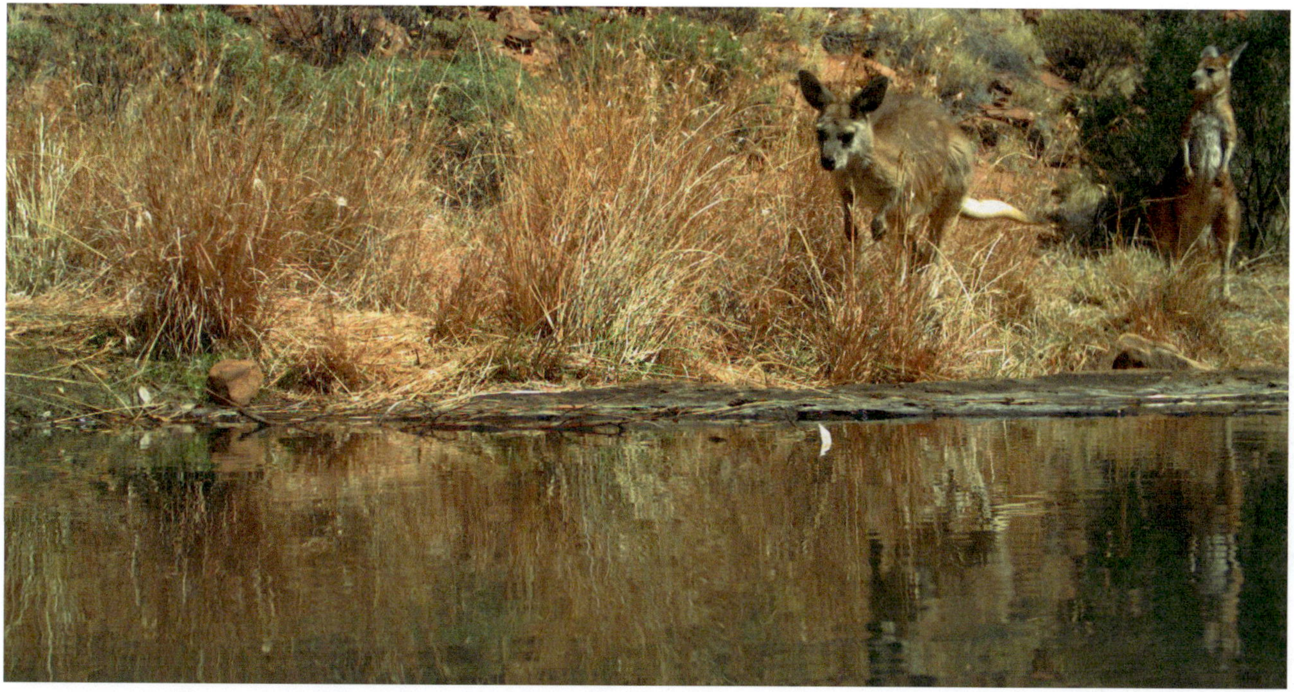

The camera at the intermittent pool has caught a youngster in the foreground and an older euro in the background.

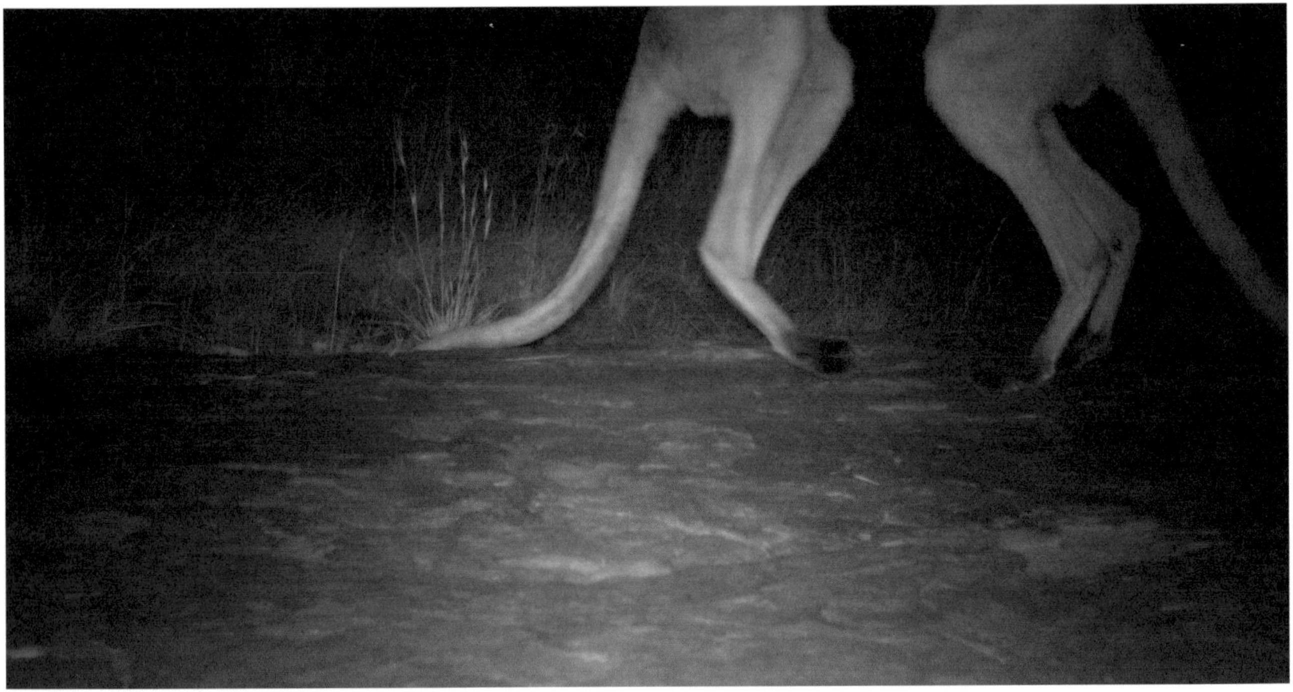

Although this is not the complete picture, there is enough detail here to indicate that a sparring session is in progress between two male euros.

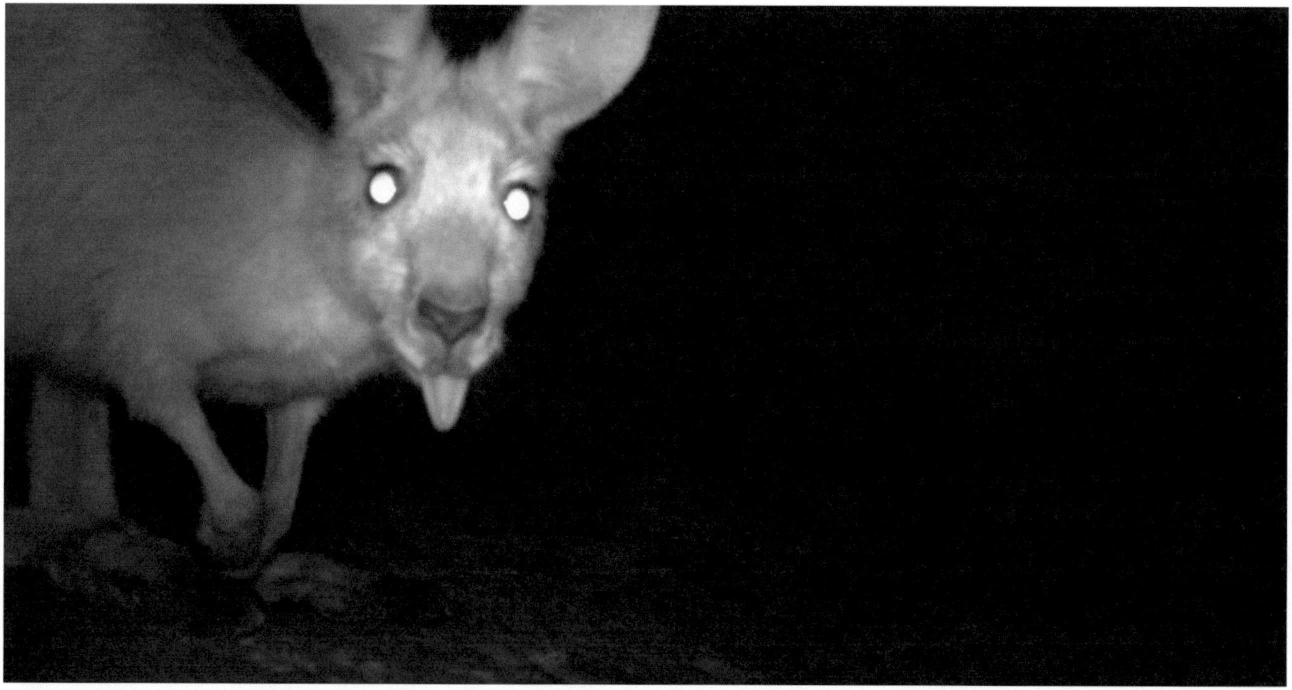

A euro selfie.

Warru, the Black-Footed Rock-Wallaby, *Petrogale lateralis centralis*

Rock-wallabies, as their name suggests, occupy rocky habitats and often occur in the steepest and most rugged of terrains. They occur in the Park but they are rarely seen during the day. They are well adapted to arid environments and conserve water by sheltering within rock crevices and rocky overhangs during daylight hours. Few images of this species have been recorded on the waterhole cameras and these are always at night. They are thought to rarely drink: however, the Watarrka camera traps have recorded individuals drinking at waterholes. Populations of this species, which is patchily distributed across central Australia, have declined and threats include predation by feral cats and introduced foxes, competition from introduced herbivores and over-abundant native herbivores, changed fire regimes and loss of genetic diversity (Lavery *et al.* 2021). Given the ongoing presence of feral cats in the Park recorded by the cameras it is possible that they are a major local predator of rock-wallabies. The presence of large numbers of a native marsupial herbivore, the euro, in the Park may also result in significant competition for nutritious plants in upland areas when food resources are limited during times of drought. Feral goats (*Capra hercus*) which in other regions are a recognised threat to rock-wallabies, through competition as an introduced herbivore, are not present in the Park.

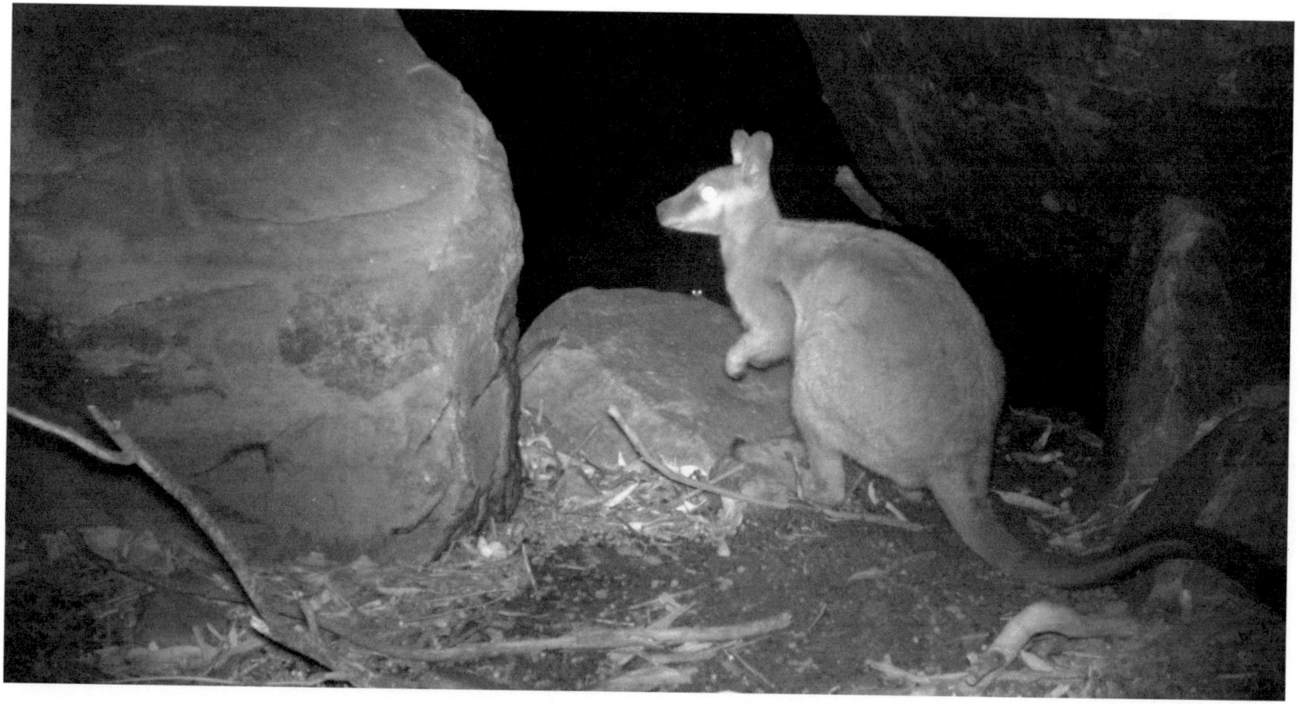

A warru at the perennial rockhole.

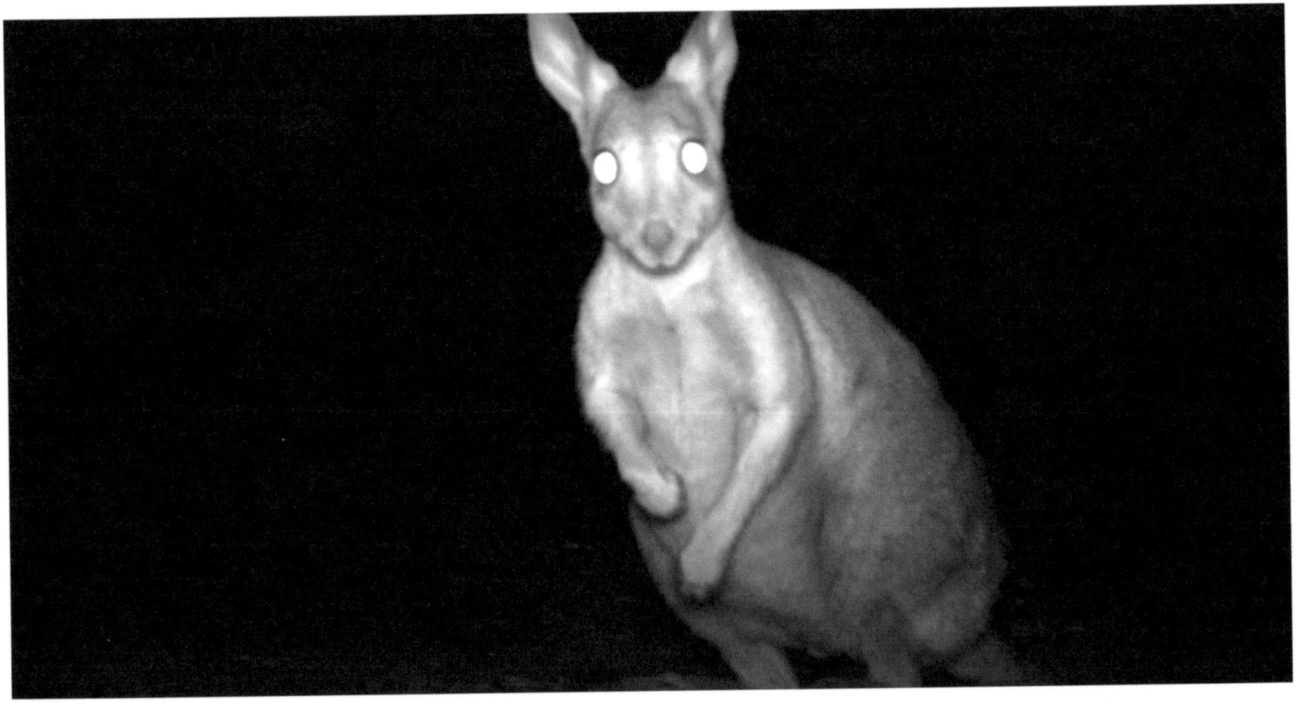

Social media has declared that the Western Australian quokka is the happiest animal on the planet. This smile suggests that the warru might be the second happiest.

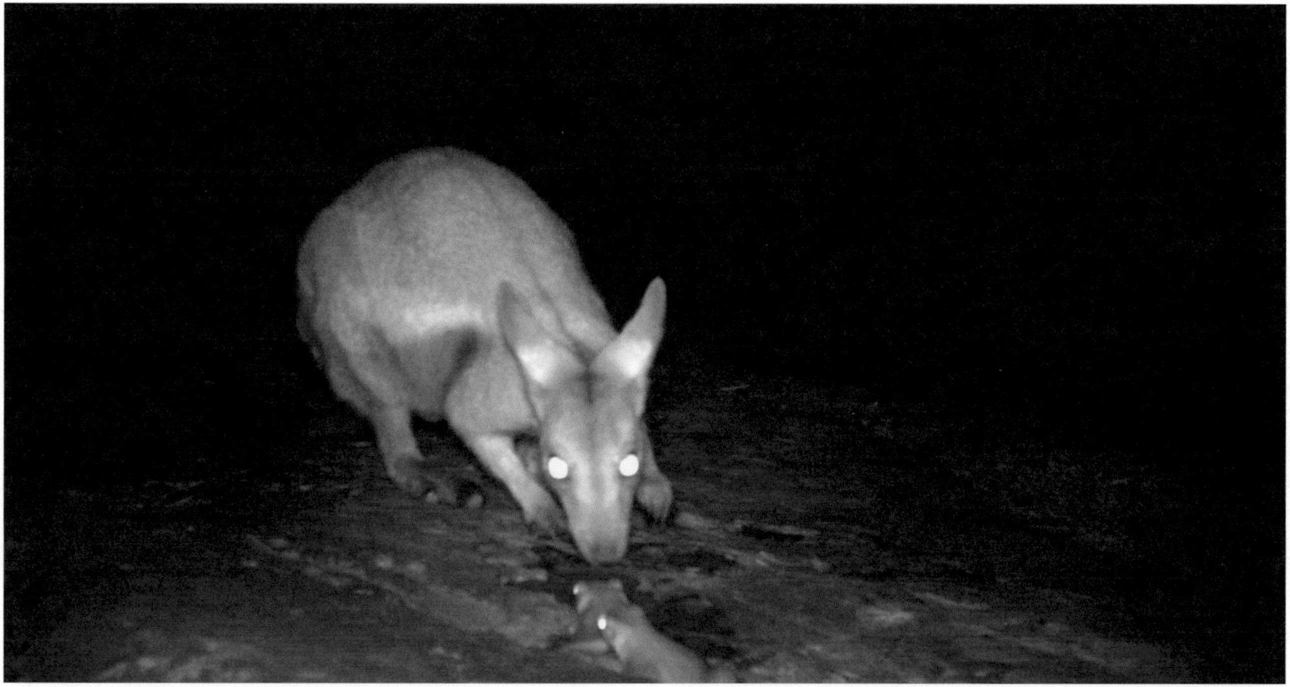

That moment when species meet - in this case the warru is drinking from a freshwater seep occupied by two Centralian tree frogs.

Fat-Tailed Pseudantechinus, *Pseudantechinus macdonnellensis*

This species occurs in the rocky ranges of central Australia, but it is not normally associated with waterholes. The soft sands and bare rocks probably make it easier to move around this part of the Watarrka escarpment. The stream banks may also provide rocky crevices for sheltering during the day with nearby basking sites. As the name suggests, this species has a distinctively tapering fat tail. It is here that fat is stored to aid survival during times of food scarcity. It is a carnivorous species, feeding on terrestrial insects including termites, beetles, grasshoppers and spiders. It is also a likely prey species for a range of predators recorded by the camera traps, including perenties, Centralian carpet pythons, black-headed monitors, various raptors, dingoes and feral cats. Studies of the ecology of this species in Tjoritja/West MacDonnell National Park by Gilfillan (2001a&b) found that they do very well in the rocky habitats of the range country. Their population numbers are more stable compared to related species that live on the adjacent arid plains. The latter show marked fluctuations in numbers (irruptions) in response to rainfall. The rocky range country likely provides a more constant source of invertebrate food species throughout the seasons.

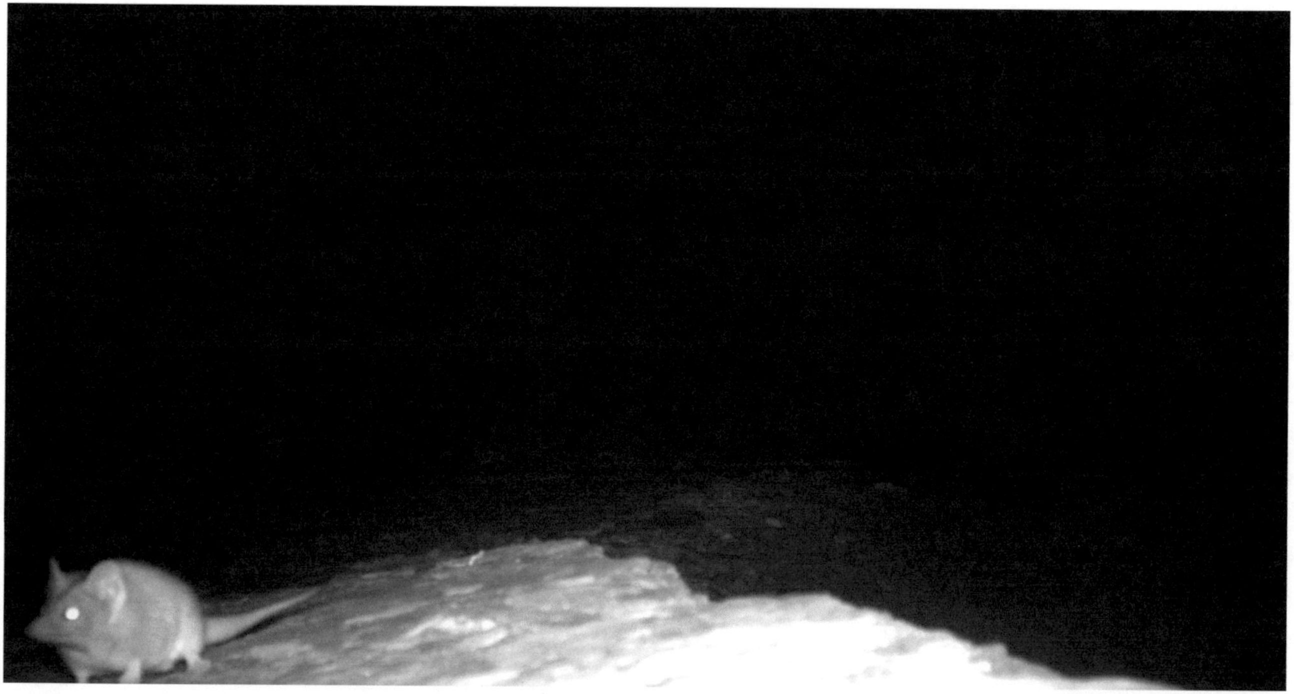

A rare event — an image of fat-tailed pseudantechinus caught on the camera at the dry temporary stream pool.

Rodents

Sandy Inland Mouse, *Pseudomys hermannsburgensis*

Although it is difficult to differentiate the sandy inland mouse from several other species on the basis of black and white camera trap images, the consensus of expert opinion on this image suggests that it depicts this species, which is distributed widely, although sparsely, across arid and semi-arid Australia. The sandy inland mouse is not considered to be a waterhole specialist, but images regularly captured by the spring camera reveal that at least one, and probably several members of this species, have made this perennial spring their home since 2014. As an omnivore it is likely to have a plentiful supply of invertebrates, including spiders, beetles and wasps, at this spring. The dense vegetation at this spring, particularly the dense fern banks and overhanging cycads, provides good habitat for both this species and its prey. The species is nocturnal and has only ever appeared on the camera at night. This nocturnal activity appears to be a very safe strategy, given that predatory species, including dingoes, perenties, black-headed monitors, brown goshawks, collared sparrowhawks, brown falcons and spotted harriers, have all been recorded visiting this spring during the day.

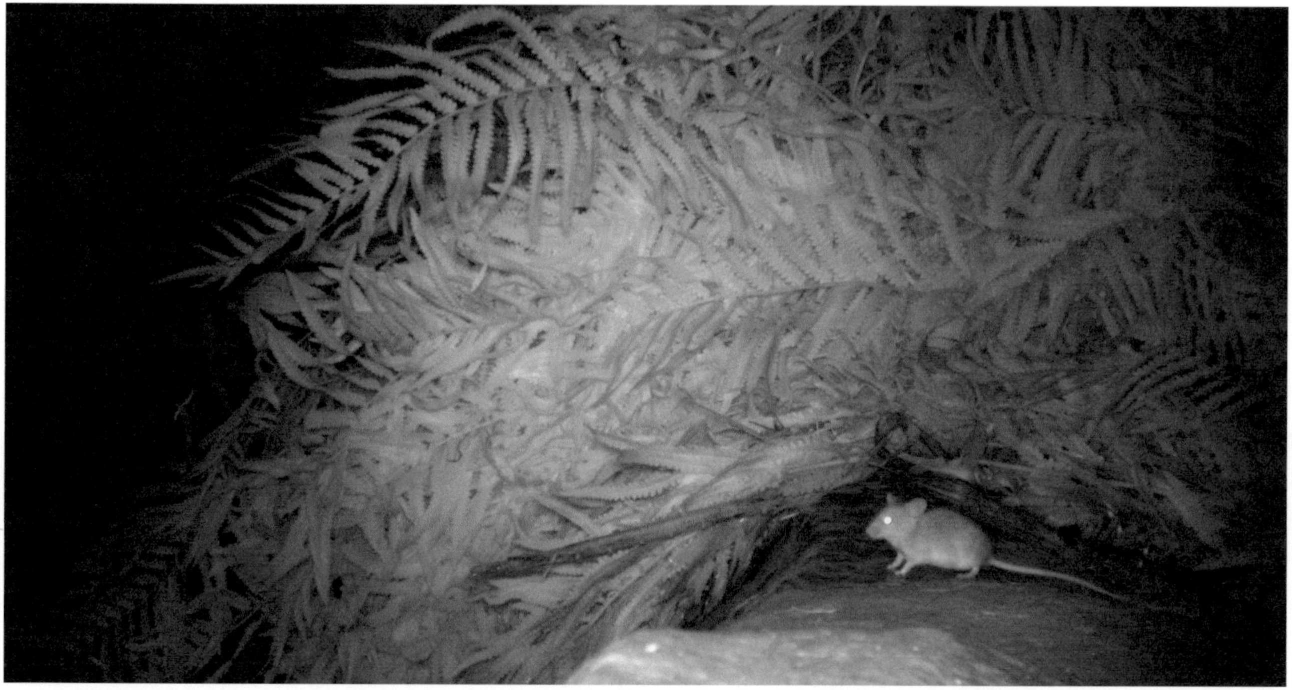

A sandy inland mouse on the edge of the perennial spring.

Raptors

The Park supports many avian predators (raptors) at the top of local food webs. The wedge-tailed eagle/walawuru, *Aquila audax*, is the largest bird of prey in Australia. It is often seen gliding high above the Park but it is also regularly recorded on the waterhole cameras. Collared sparrowhawks and brown goshawks often visit the perennial spring where the zebra finches that arrive almost continually to drink provide an abundant source of food. The camera traps have revealed how important permanent water is for raptors during hot weather, as they need water for cooling as well as drinking.

Simon Votto, in an Honours project at the University of Canberra, used the camera trap dataset to investigate how often birds came to drink at the waterholes. He found that a temperature of 35°C is critical. Once this threshold is reached, more raptors come to the spring and they stay for longer. This excludes the smaller birds they prey on, such as zebra finches and diamond doves. Understandably, these birds do not come into the spring to drink while a raptor is present. This finding has climate change implications because it suggests that as temperatures increase, and droughts and heatwaves become more common, the competition for access to permanent water will become intense. Under this scenario, the probability that local extinctions will occur is high (Votto *et al.* 2020).

Wedge-Tailed Eagle/Walawuru, *Aquila audax*

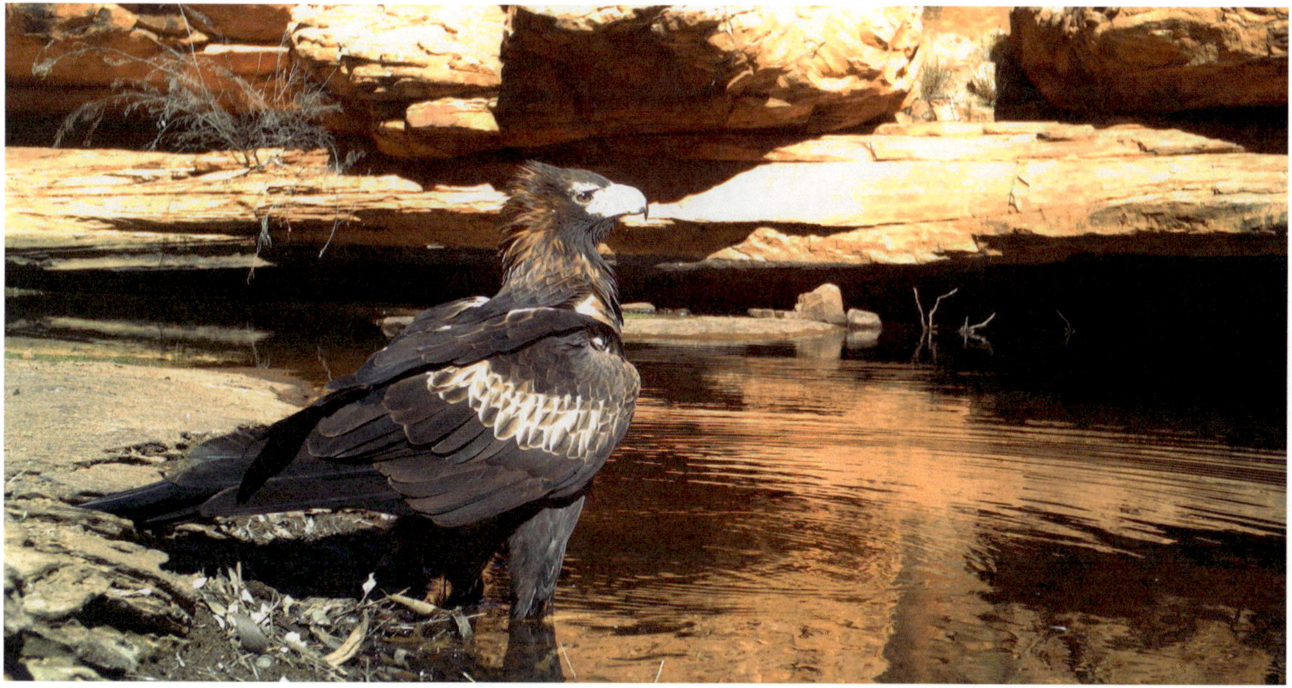

Wedge-tailed eagles / walawuru visit waterholes in the Park to drink and bathe.

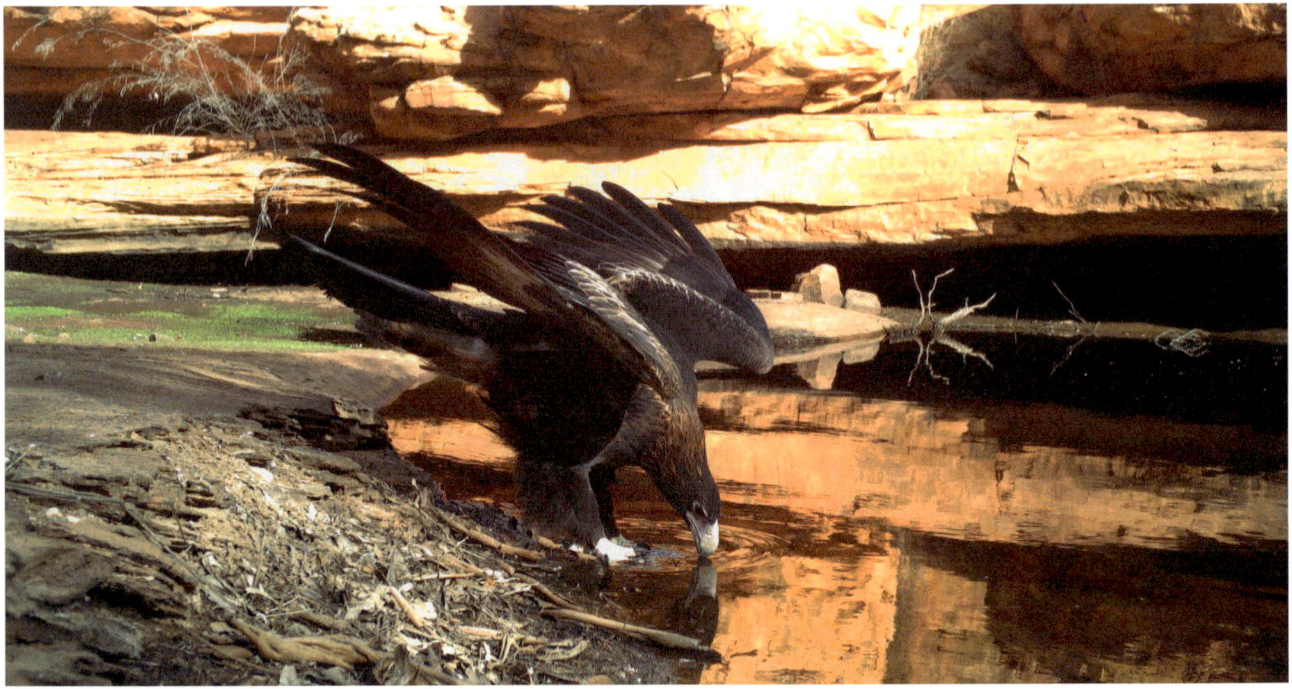

The size and large wingspan of wedge-tailed eagles means that they can access all parts of the Park.

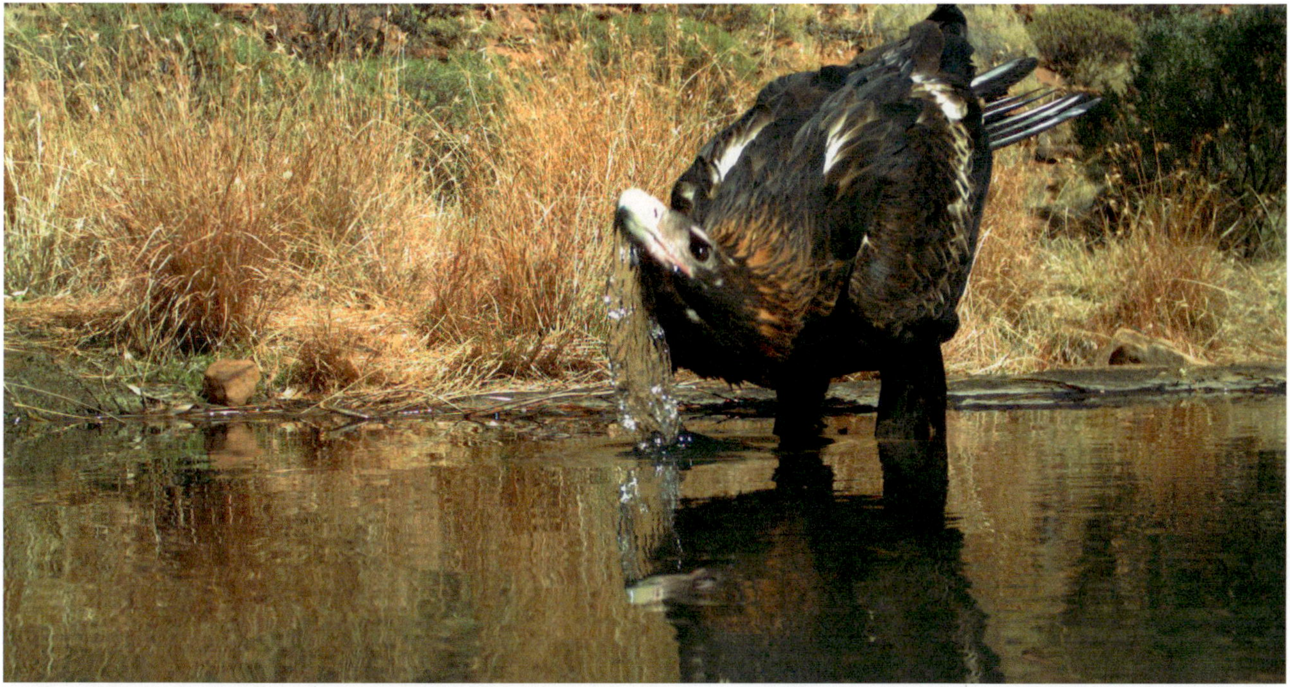

Although it is likely that wedge-tailed eagles can obtain most of their water requirements from their prey, they still often drink from Park waterholes.

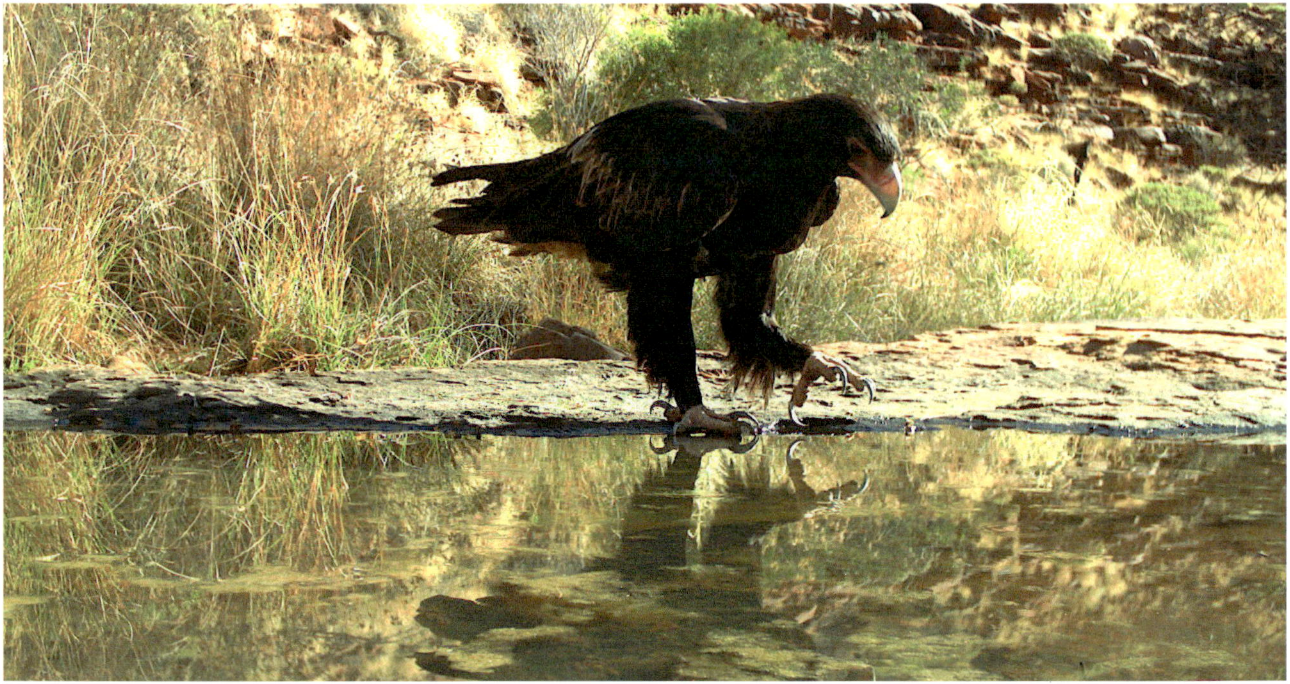

The impressive talons of the walawuru are displayed here.

Little Eagle, *Hieraaetus morphnoides*

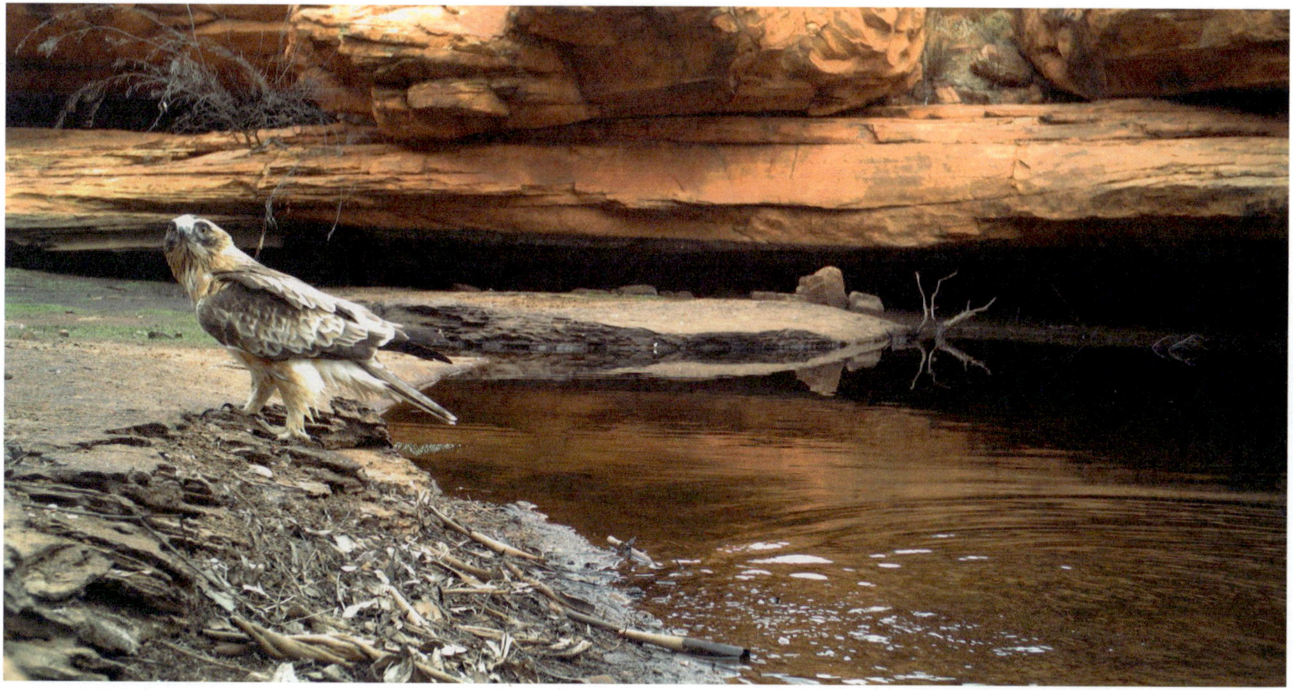

The little eagle is more likely to be seen flying over nearby woodlands than at the waterholes. It is smaller than a wedge-tailed eagle and distinguished by its highly feathered legs (trousers).

Black-Breasted Buzzard/Aralapalpalpa, *Hamirostra melanosternon*

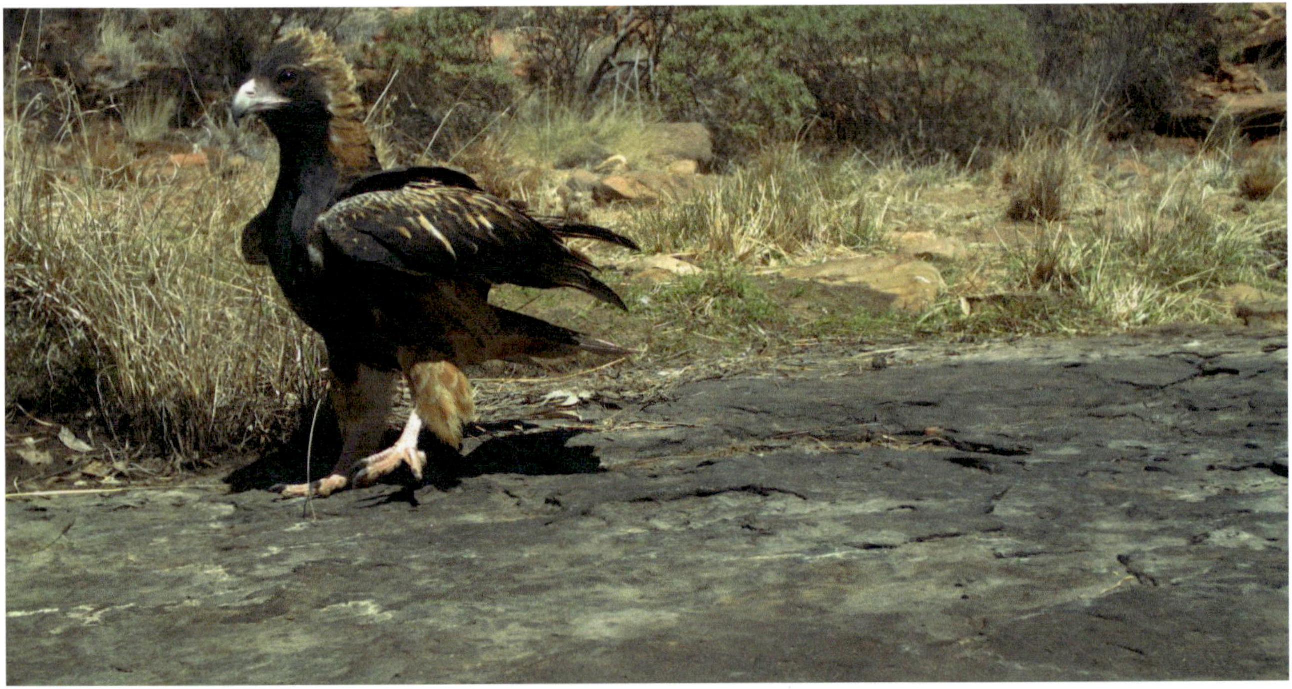

A rare capture on the intermittent stream pool camera, a black-breasted buzzard/aralapalpalpa. This species has the special skill of cracking eggs by various means, including throwing stones at them (Auman 1990).

Brown Falcon/Kirkinpa, *Falco berigora*

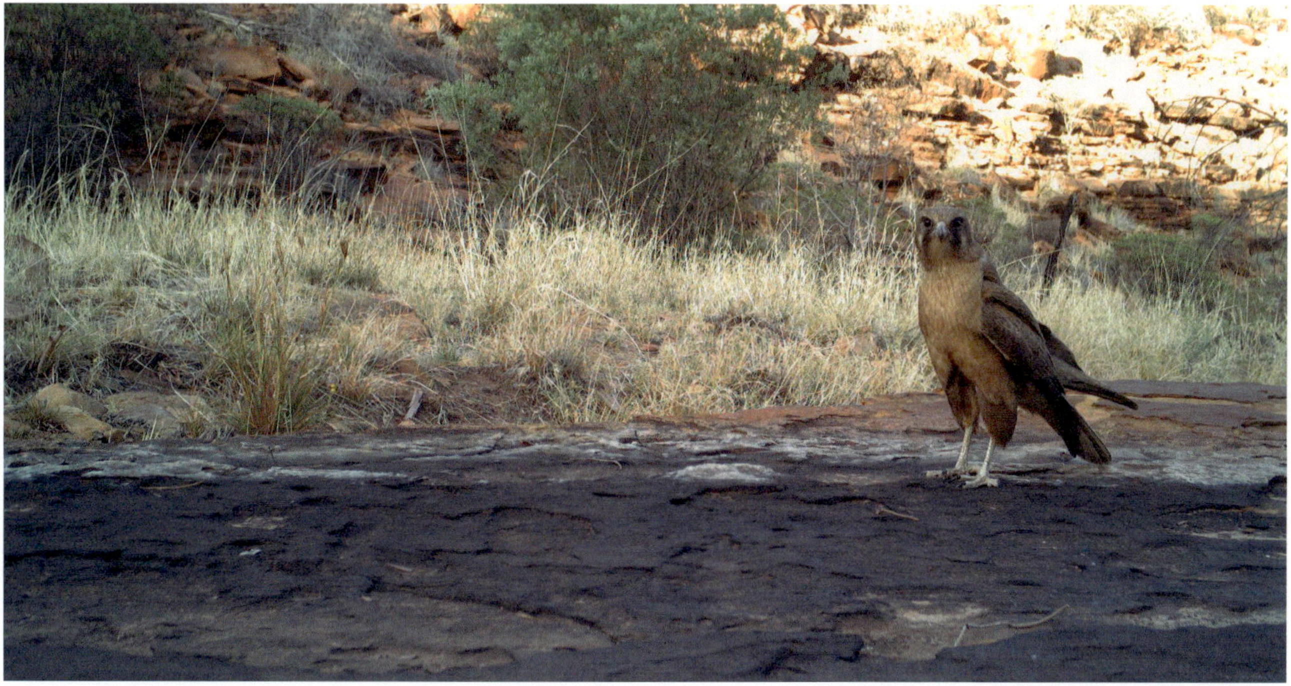

The brown falcon/kirkinpa, is commonly seen perched high on dead trees or power poles. This image is one of very few images of the falcon on the ground.

Collared Sparrowhawk, *Accipiter cirrocephalus*, and Brown Goshawk, *Accipiter fasciatus*

Collared sparrowhawks and brown goshawks are ambush predators. They often appear at the permanent waterholes to both drink and bathe. The dense vegetation fringing the waterholes is also a good place to hunt prey. The two species can be hard to tell apart. The distinguishing features of the collared sparrowhawk include its long thin legs with long 'trousers', a long middle toe and a slightly surprised expression. In contrast, brown goshawks are usually larger, with thicker legs, rounded tail feathers, flatter head, and a fiercer stare.

Although the presence of both species at the same time at a waterhole is rare, images recorded at the perennial spring indicate that a brown goshawk and a collared sparrowhawk, visited the spring together during the very hot and dry conditions that persisted throughout the summer of 2019.

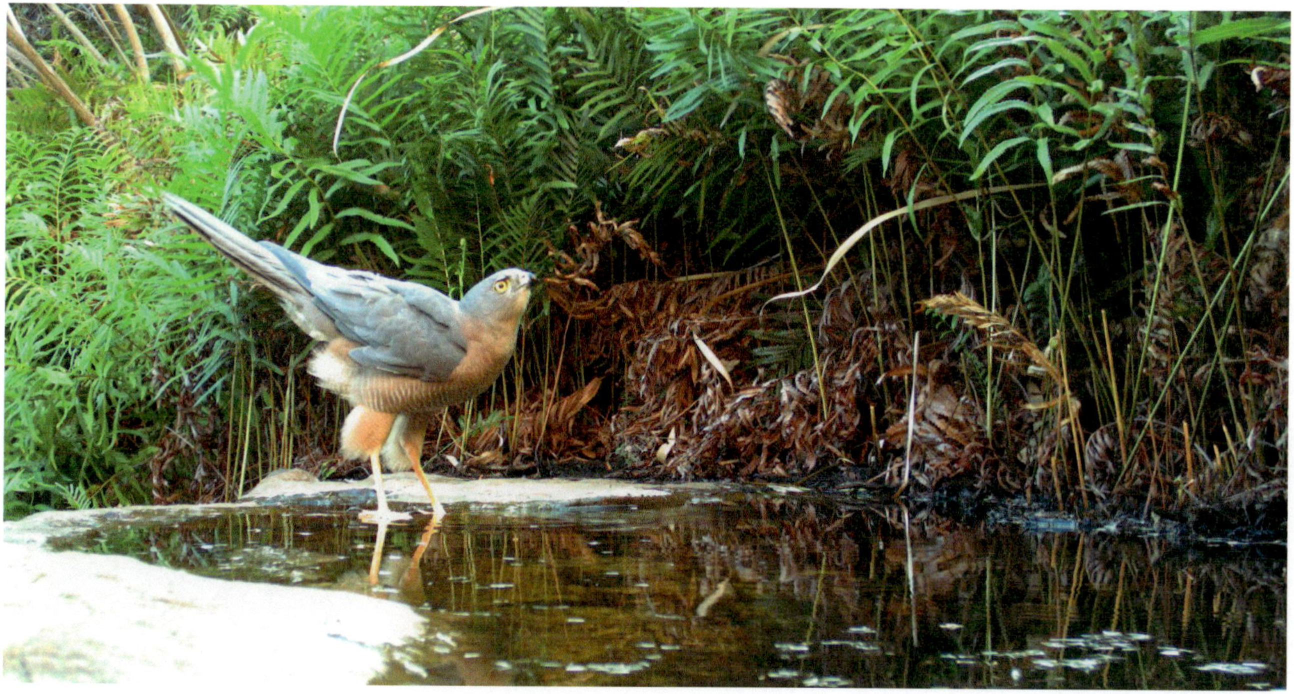

This collared sparrowhawk is a regular visitor to the perennial spring pool. Its long 'trousers' and thin legs distinguish this species from the brown goshawk.

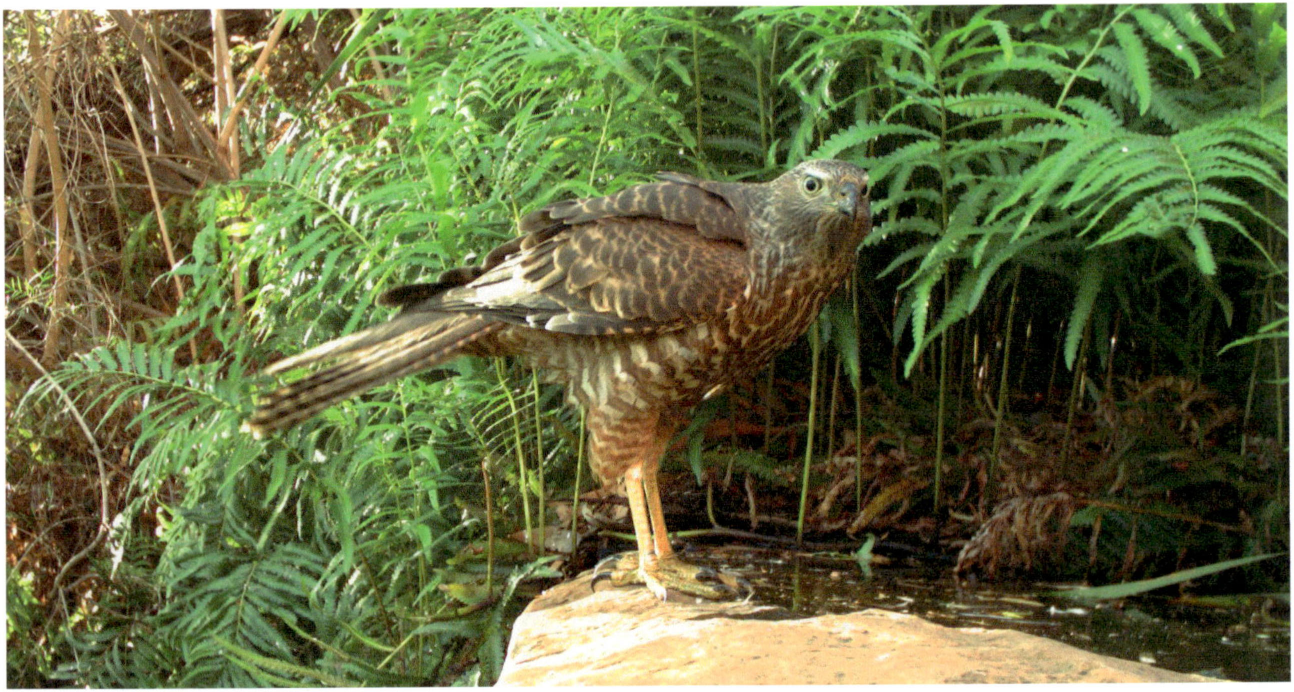

A brown goshawk is also a frequent visitor at the perennial spring pool on hot days.

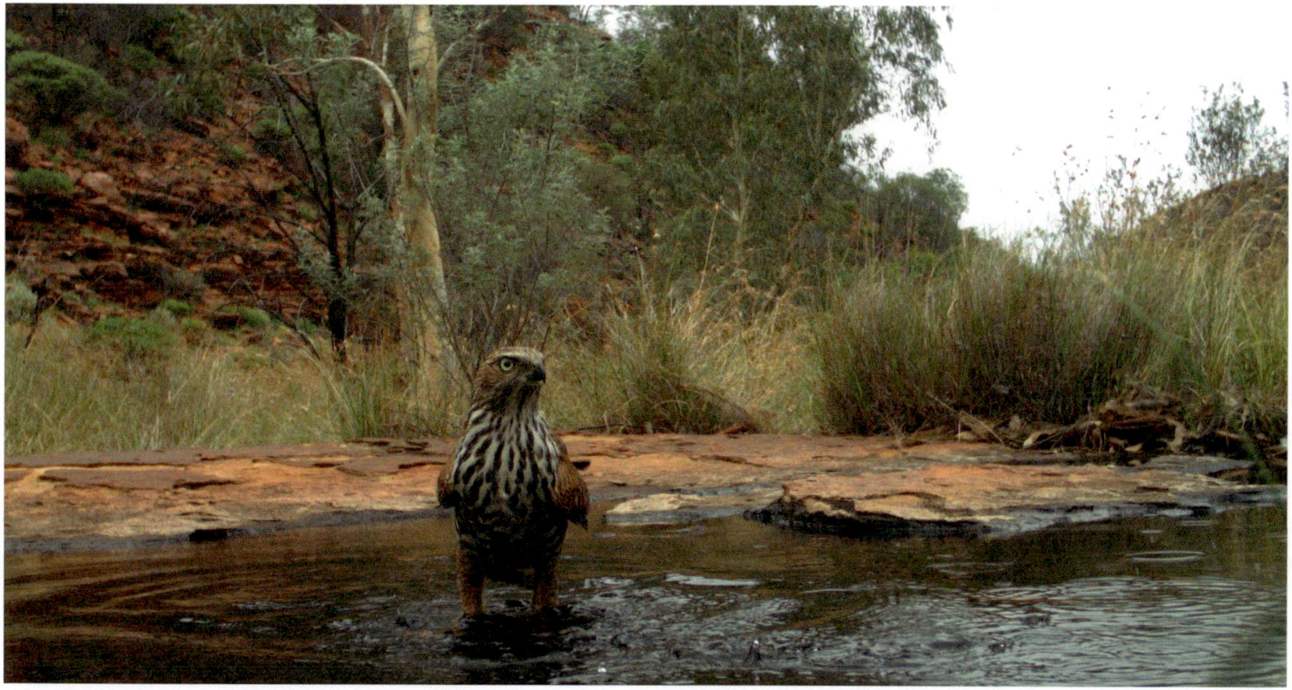

A juvenile brown goshawk at the intermittent stream pool after it filled from a summer rainstorm.

Omnivorous Birds

As the name suggests, omnivores have a diet that includes a wide range of prey items. They may be considered less likely to visit waterholes because their diet often includes prey with a high water content. However, many omnivores are regularly recorded on the waterhole cameras. This is probably because the moist areas fringing the waterholes support more plants than the surrounding arid landscapes. These plants provide habitats for small vertebrates and invertebrates, and are sources of seeds and fruits. All of these are food items for omnivorous birds.

Some of the omnivorous birds that are most commonly recorded at the Watarrka waterholes include the western bowerbird/ikarka, the Torresian crow/kaanka, the grey shrike-thrush/wititata, and the willie wagtail/tjintir-tjintir. These are the birds that you are often likely to see, or hear, within the Park. Not surprisingly, their wide-ranging diets means that they may move around more freely, rather than being restricted to specific habitats.

Western Bowerbird/Ikarka, ***Ptilonorhynchus guttatus***

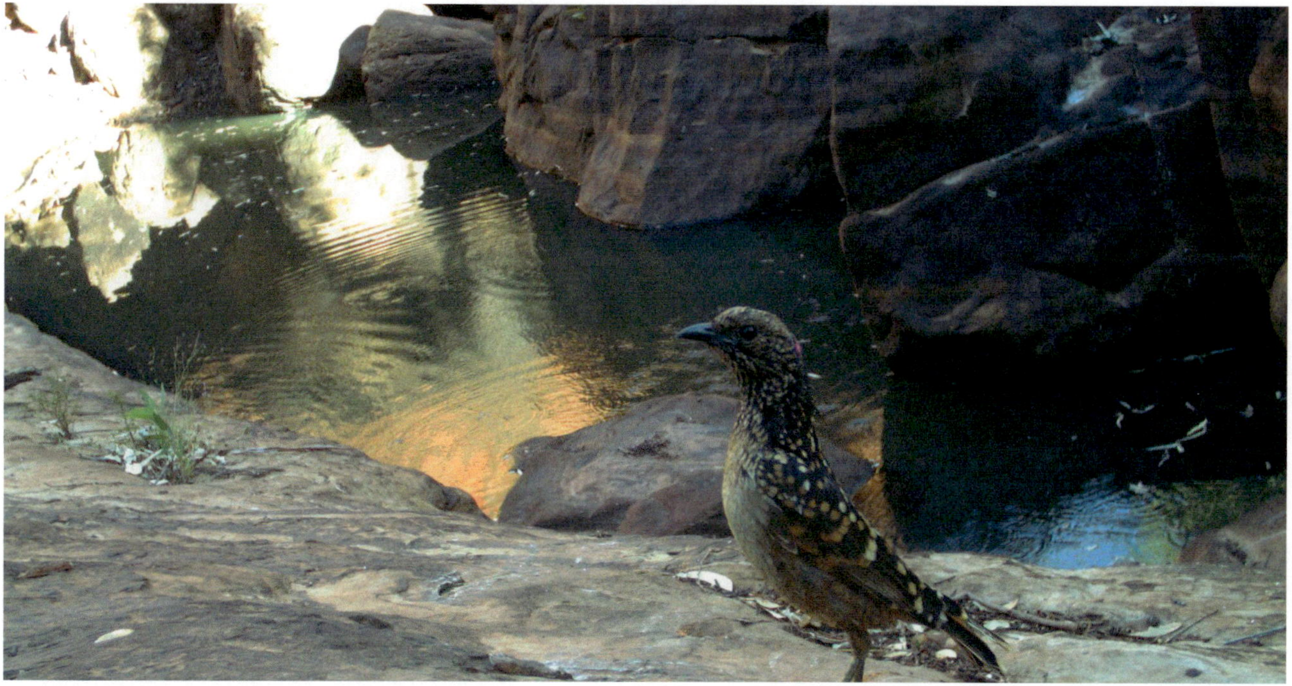

The western bowerbird/ikarka, is an almost daily visitor at the two permanent water camera trap sites. The male's distinctive bower, comprising treasured white and green objects, is usually located within thick bushes nearby.

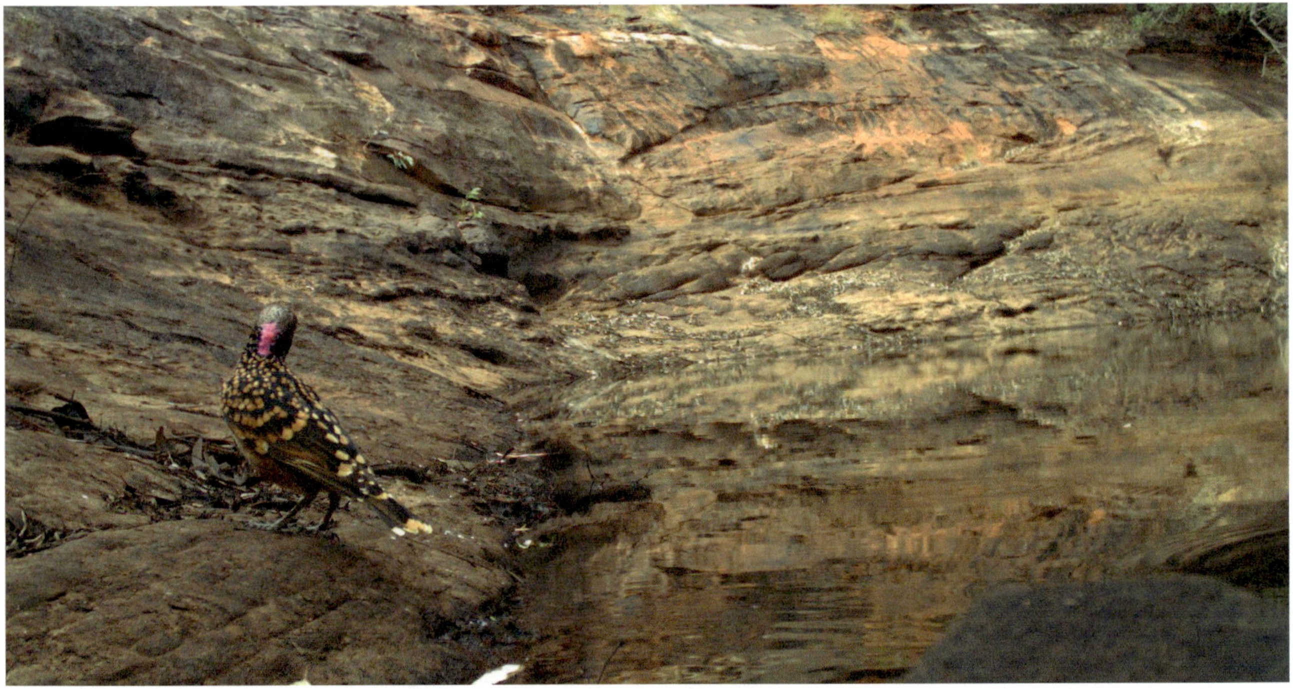

Apart from that bright pink neck patch, the colours of this male western bowerbird provide excellent camouflage because they closely match the colours of the rocks surrounding the waterhole.

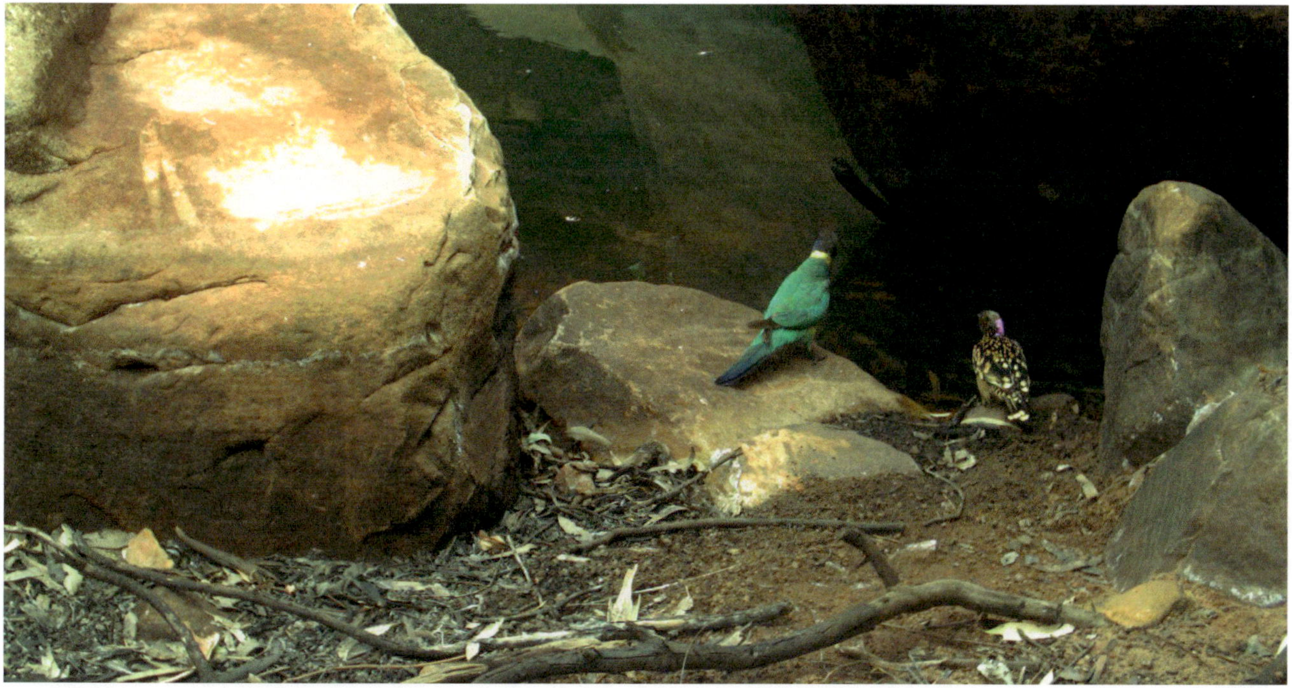

Many species share the waterholes, in this case, a western bowerbird and an Australian ringneck parrot/patilpa, *Barnardius zonarius*, were captured together on camera.

Torresian Crow/Kaanka, *Corvus orru*

Crows are the ultimate opportunists, they are both predators and scavengers. Torresian crows/kaanka are common in the Park and the waterhole camera images provide ample evidence that they are intelligent, curious and mischievous birds. The cameras are often the target of interest and sometimes vandalism. Crows will turn cameras over if they are not secured. Peck marks often occur on the cameras and their presence indicates how closely crows will examine a new object when it appears in an otherwise familiar location. The Indigenous name for the crow, kaanka, closely resembles the distinctive sound of a crow calling loudly across the landscape.

The Australian raven, *Corvus coronoides*, has also been recorded on the waterhole cameras, but the appearances of this species are much rarer. It likely prefers the more open habitats of the nearby plains and sandhills country.

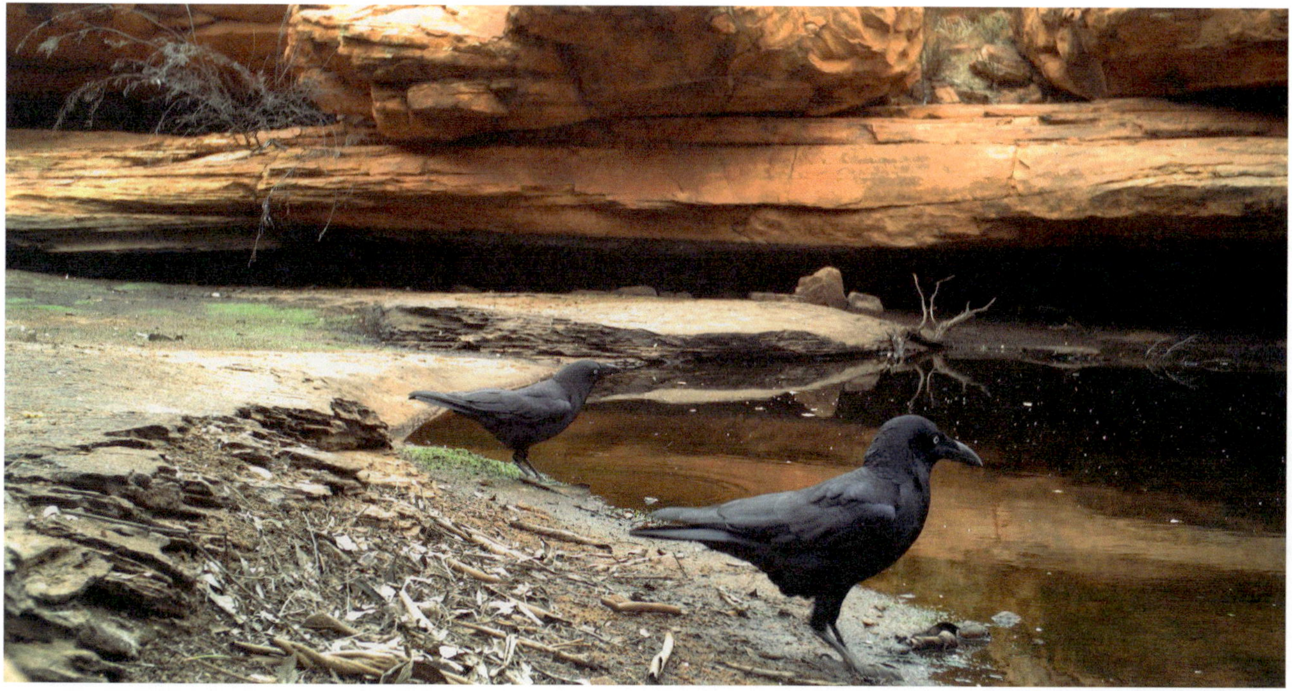

Torresian crows often appear at the waterholes in pairs.

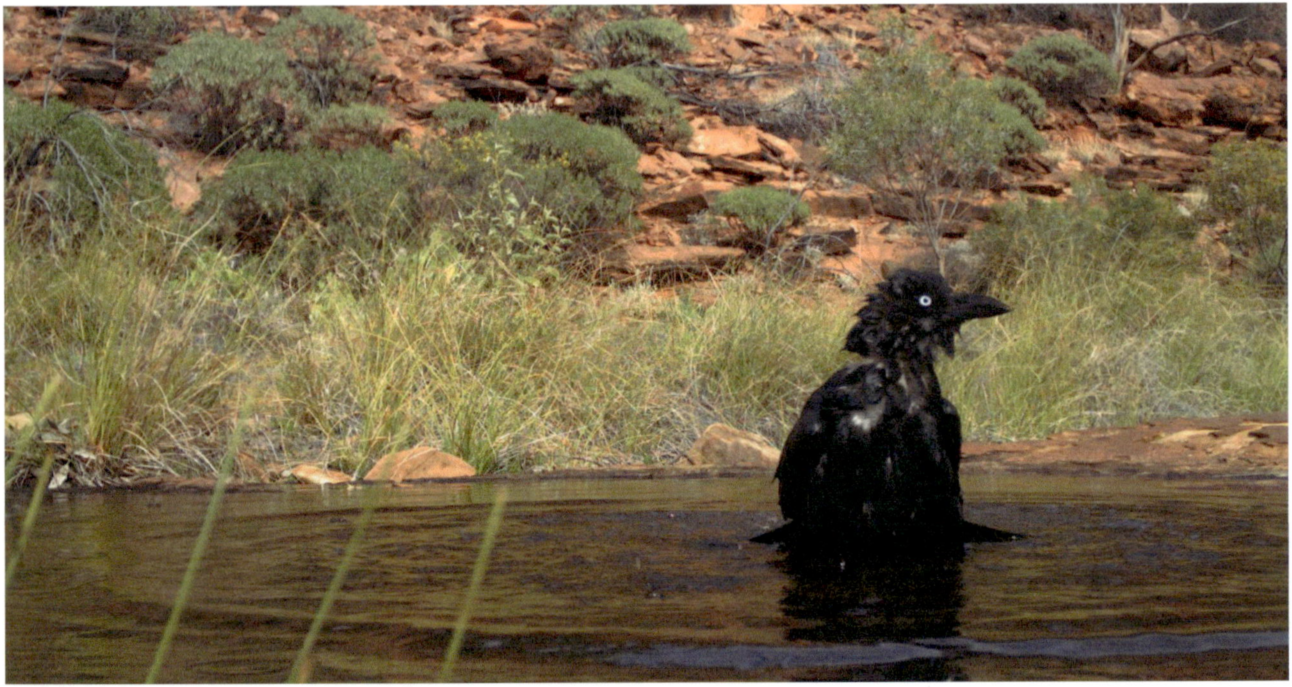

Like something out of a horror movie, this crow has just surfaced after a full immersion bath.

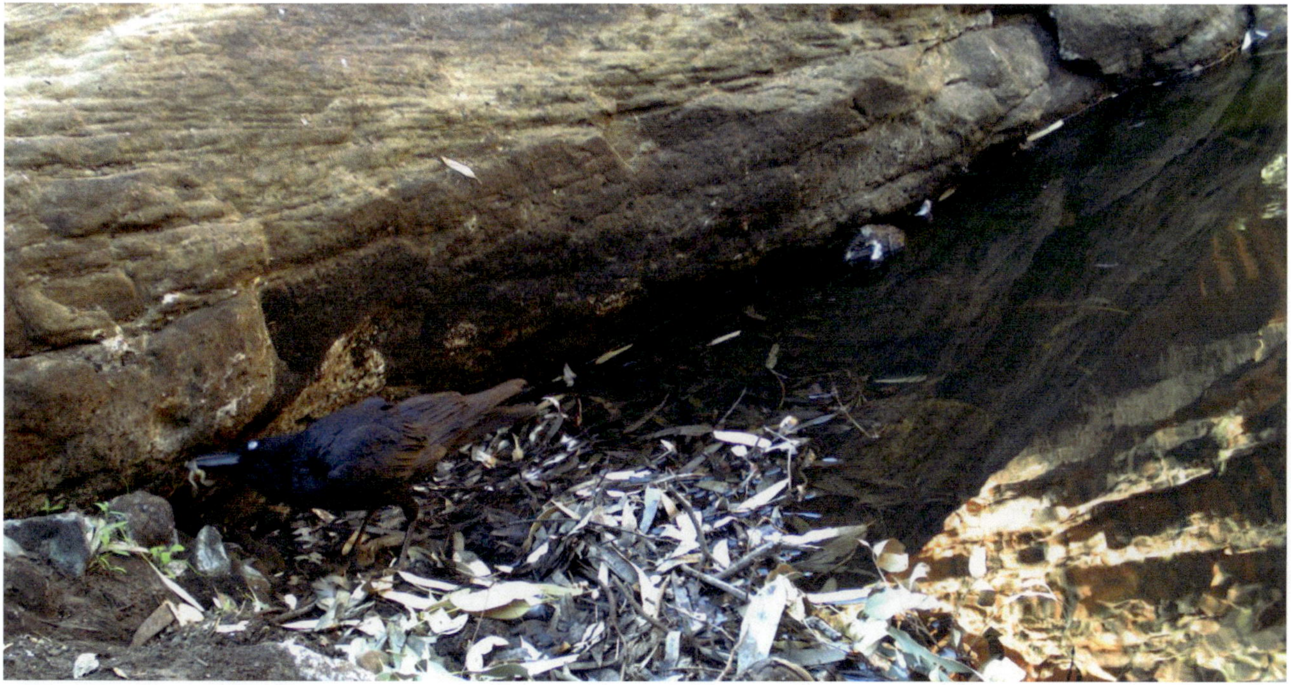

A good meal for the crow, not such a good outcome for the frog.

Grey Shrike-Thrush/Wititata, *Colluricincla harmonica*

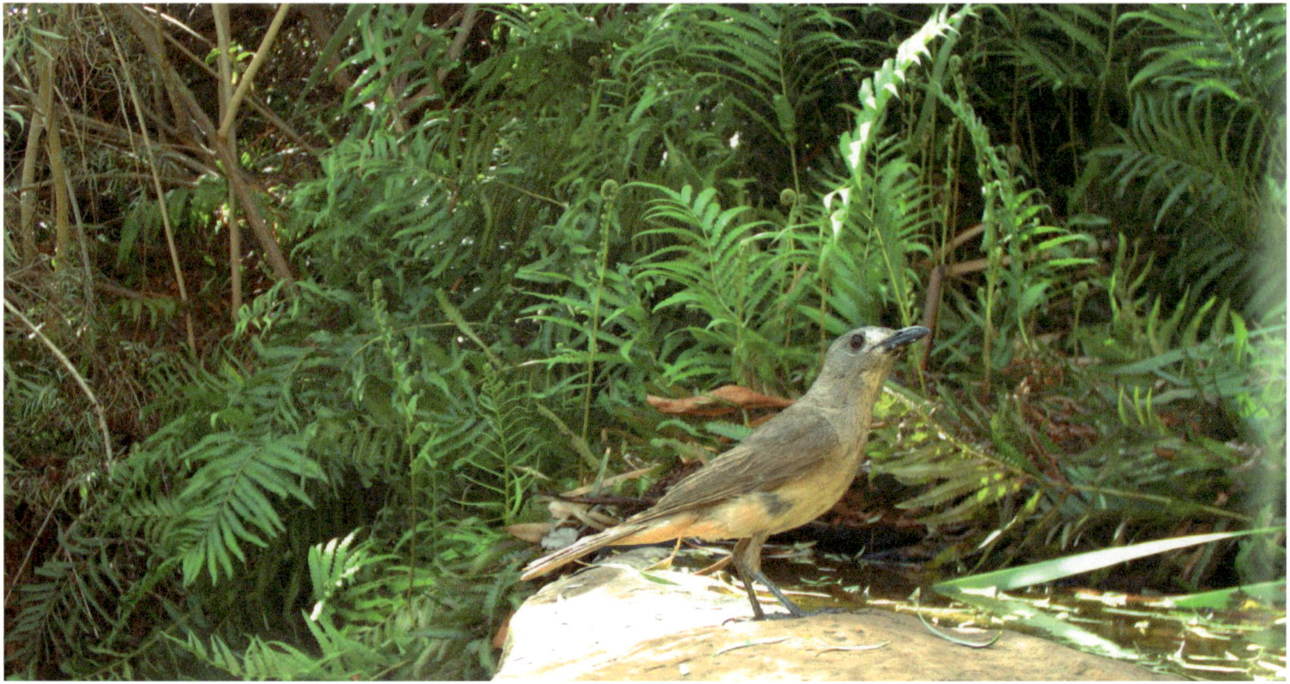

The melodic song of the grey shrike-thrush/wititata, is commonly heard throughout the Park and this species often appears on the waterhole cameras.

Willie Wagtail/Tjintir-tjintir, *Rhipidura leucophrys*

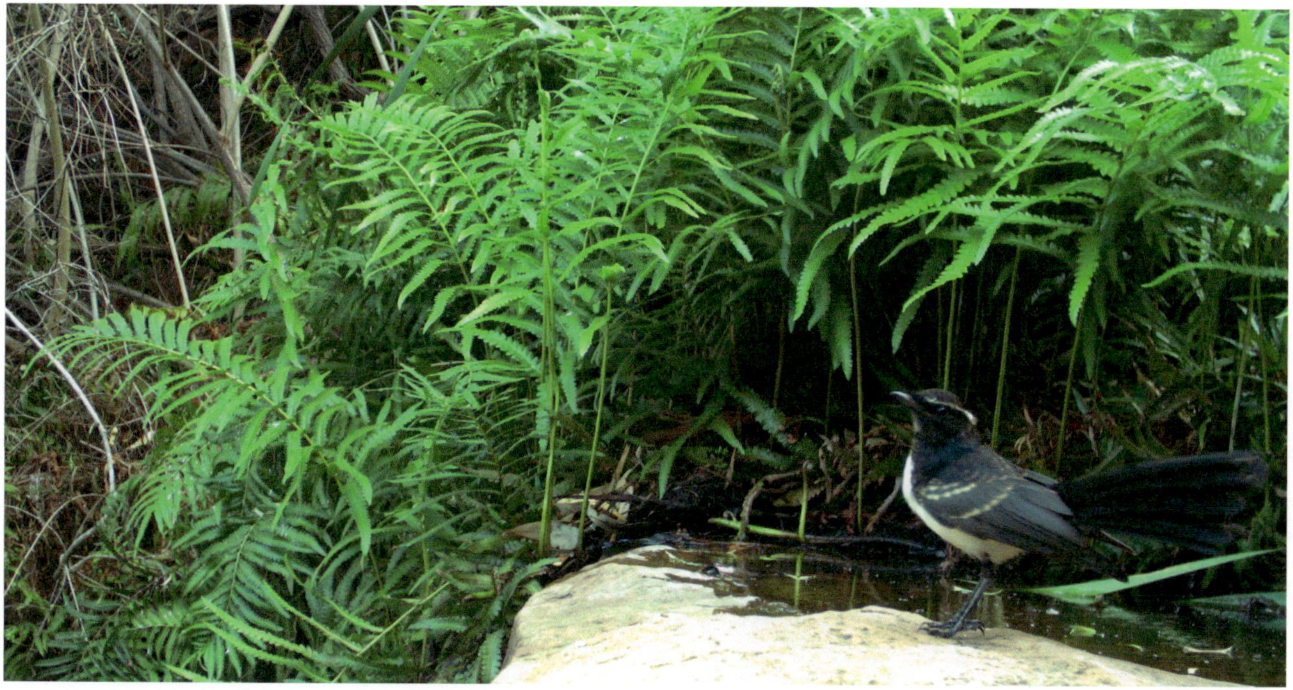

The willie wagtail is a common visitor to all the Park waterholes but its rapid movements result in images that are rarely in focus. Their Indigenous name, tjintir-tjintir, mimics the oft-repeated sound of their call.

Seed-Eating Birds

Birds that eat grains and seeds (granivores) find much to eat in the open grasslands and spinifex-dominated ranges of inland Australia. However, their dry diet means that they always need to be close to a source of water. Zebra finches, *Taeniopygia guttata,* are the most commonly recorded granivores at permanent waterholes. The presence of this species is one of the most reliable indicators of water in arid inland Australia and hearing their distinctive soft chirping call is a sure sign that water is close by. Their Indigenous name, nyii-nyii, closely resembles the sound of their call. They are very sociable birds and usually form large flocks. This flocking behaviour acts to provide protection from predators by providing 'safety in numbers'.

If you sit at a waterhole surrounded by vegetation, you will hear the distinctive chattering of the zebra finches as they wait for you to leave before coming in to drink. If they are really thirsty, and you do not appear to be a threat, they will soon appear at the water's edge. Although many, many images have been recorded, their quick movements, small size, and large numbers, means that often part, or all, of most images are out of focus.

Zebra Finch/Nyii-nyii, *Taeniopygia guttata*

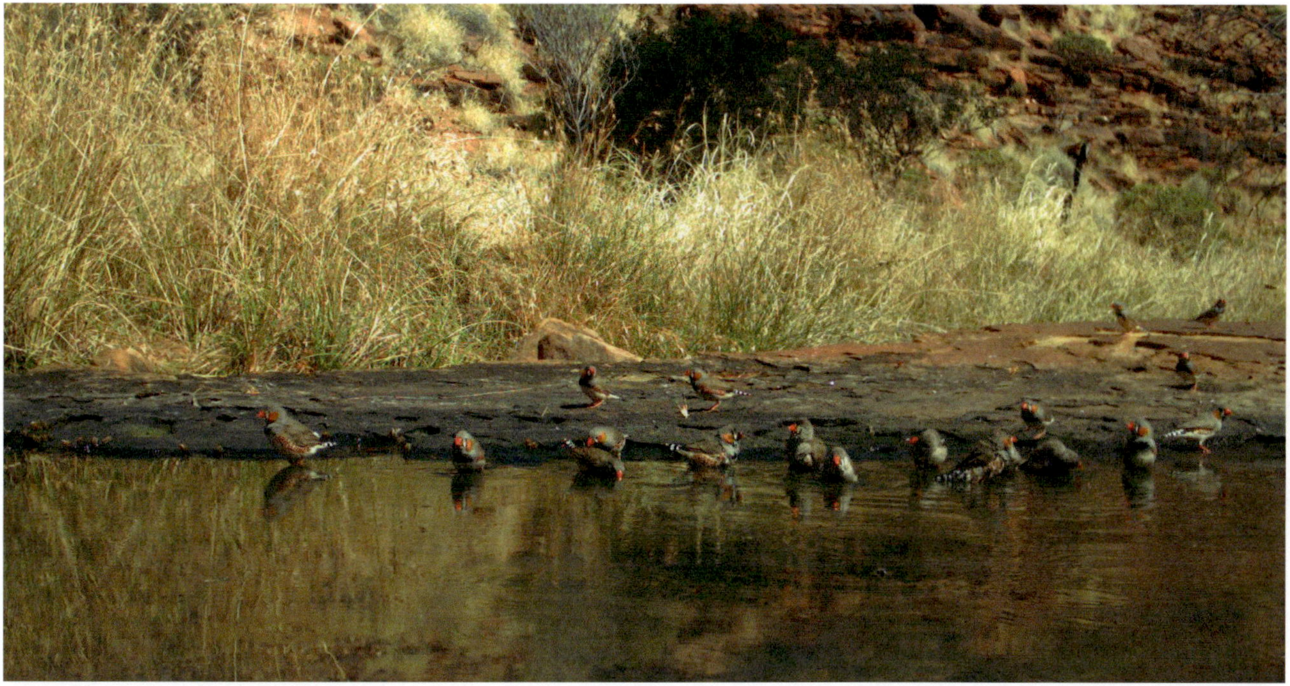

This shallow pool on the intermittent stream provides plentiful water for zebra finches and easy access from the gently sloping rocky edge.

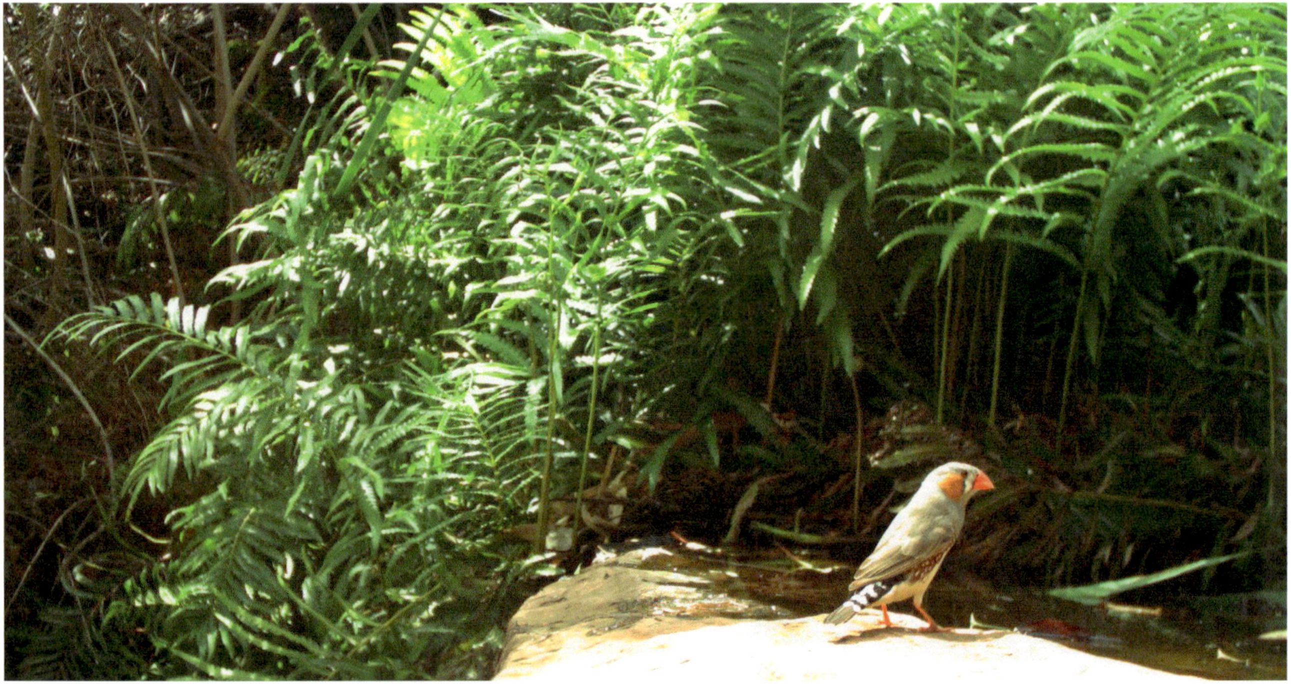

This male zebra finch appears to be alone at the perennial spring, however, other members of the flock are likely to be drinking from the pool just off camera.

Painted Finch/Ilitja, *Emblema pictum*

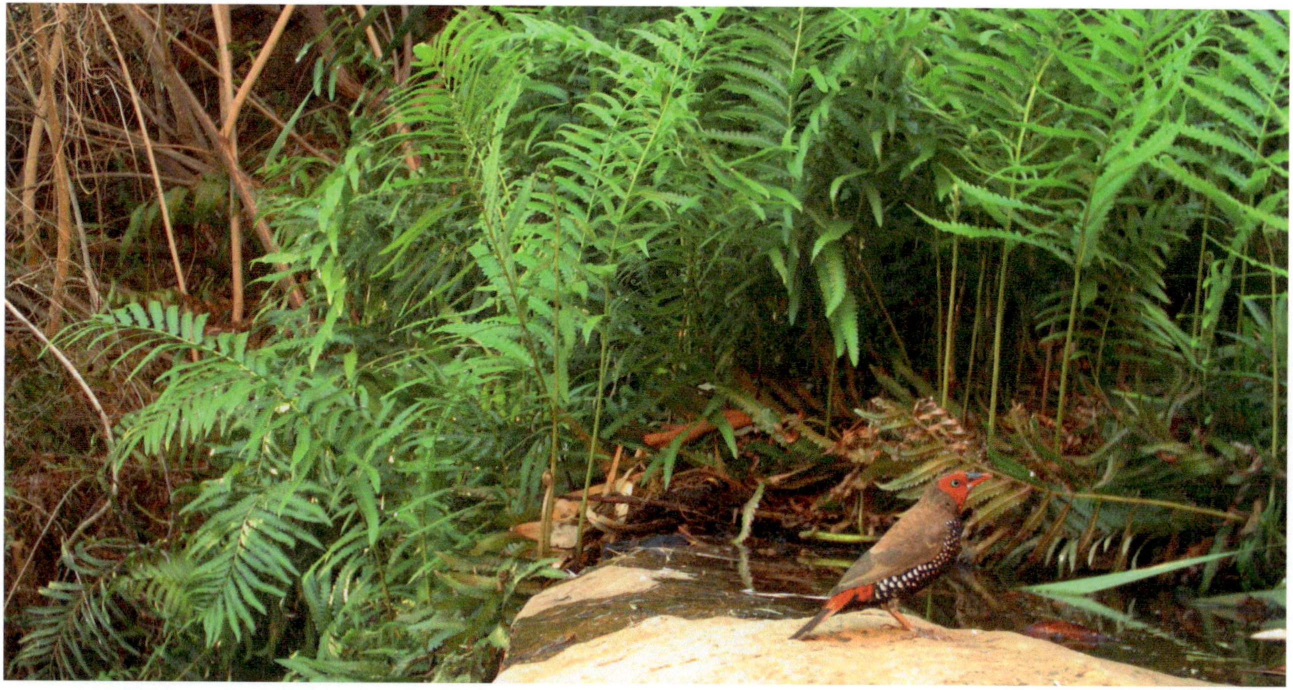

It is hard to miss the bright red head and tail colour of the male painted finch/ilitja. They often drink with zebra finches but occur in much smaller numbers.

Diamond Dove, ***Geopelia cuneata***

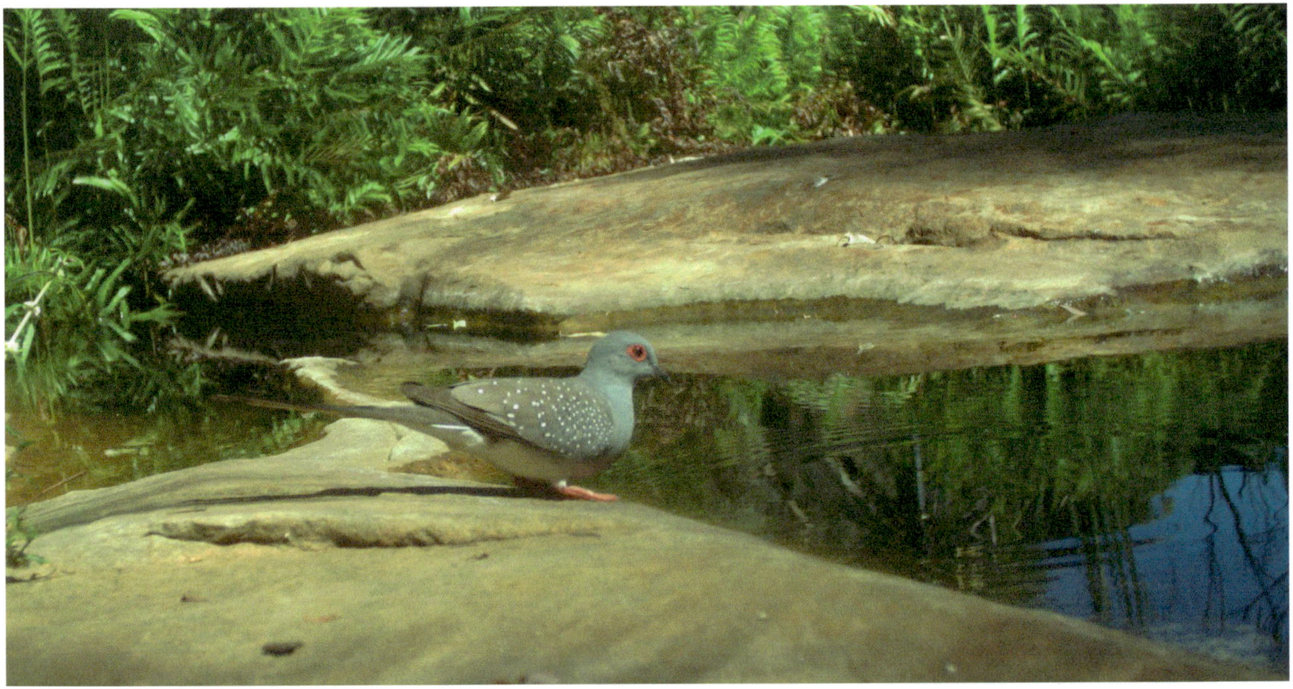

A solitary diamond dove is about to drink at the spring pool.

Spinifex Pigeon/Ipuru, *Geophaps plumifera*

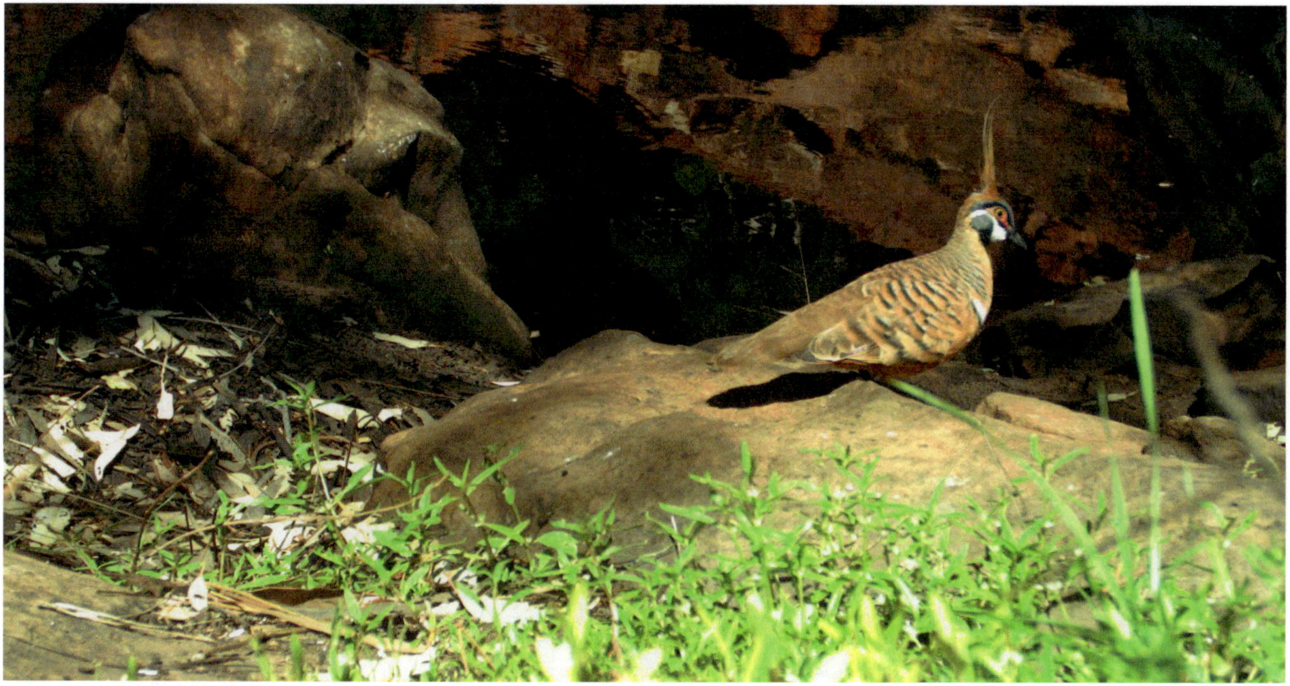

The Park supports a large population of spinifex pigeons/ipuru. The crest of the spinifex pigeon makes for good camouflage amongst clumps of spinifex and tall grasses, while the wing colours blend with the rocky and sandy surrounds.

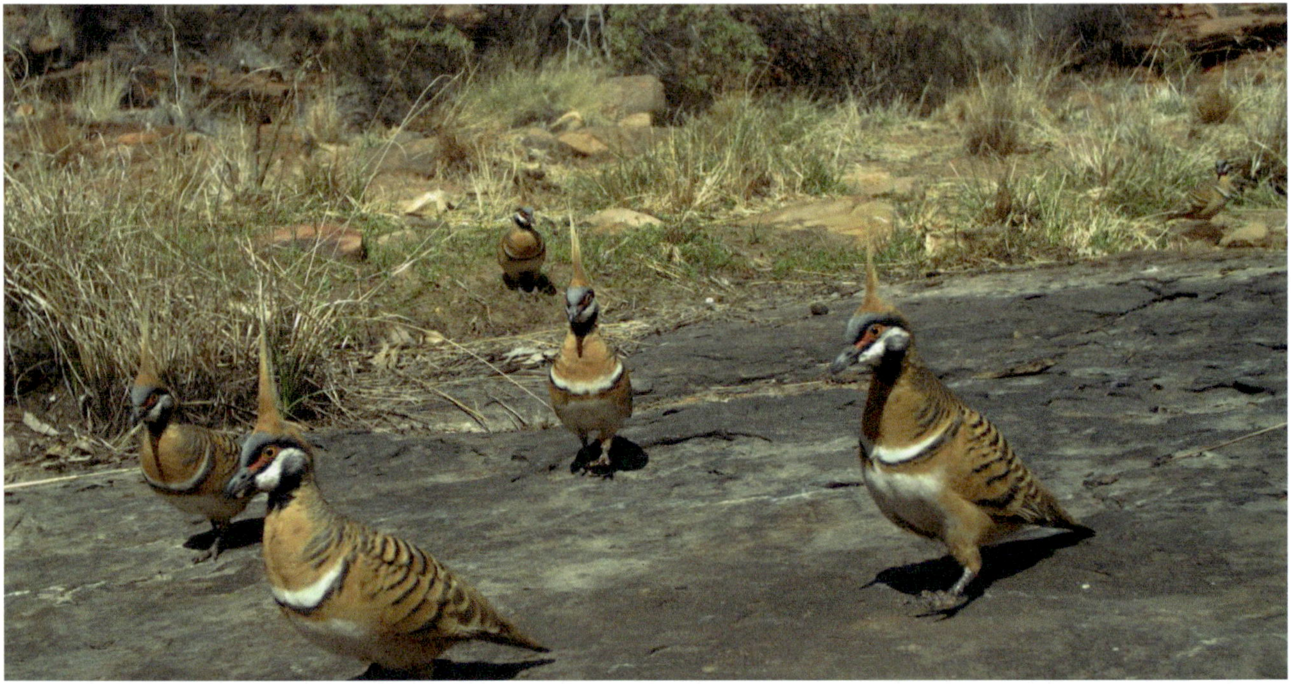

The intermittent stream pool has contracted to a puddle out of sight beneath the camera, but it still provides water for this flock of spinifex pigeons.

Bronzewing Pigeon/Manpi, ***Phaps chalcoptera***

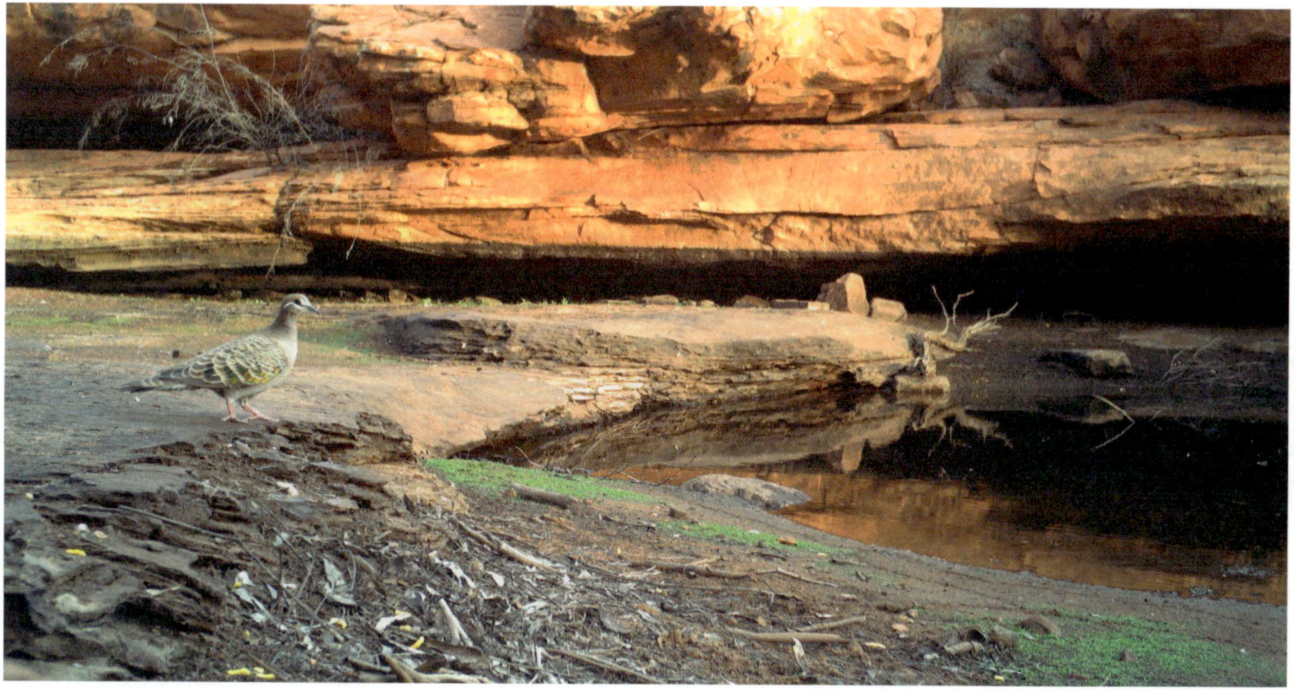

Common bronzewing pigeons/manpi, usually arrive at waterholes to drink in the late afternoon and early evening. Flocks of up to 16 pigeons have been recorded, here a solitary bronzewing was captured in the golden light of the late afternoon.

Parrots

Parrots are an iconic group with a wide distribution across the Australian continent. However, with the exception of budgerigars, they are rarely recorded on the waterhole cameras. Most species prefer the woodlands and more open grassland habitats of the Park, rather than the waterholes within the range country. In addition to the budgerigar/kilykintarri, other species that have been recorded at the waterholes include the Australian ringneck parrot/patilpa, the Major Mitchell cockatoo/kakalyalya, the mulga parrot and the cockatiel.

Parrots are predominantly seed eaters but some species will also feed on fruits and insects. Budgerigars feed almost exclusively on the dry seeds of grasses and native herbs, so they need to visit waterholes regularly to meet their daily water requirements. Budgie populations can increase very quickly after good rains have fallen in arid regions. Spectacular images have been captured of flocks of budgies drinking from Watarrka waterholes after the recent (2021) rain events.

Major Mitchell's Cockatoo/Kakalyalya, *Lophochroa leadbeateri*

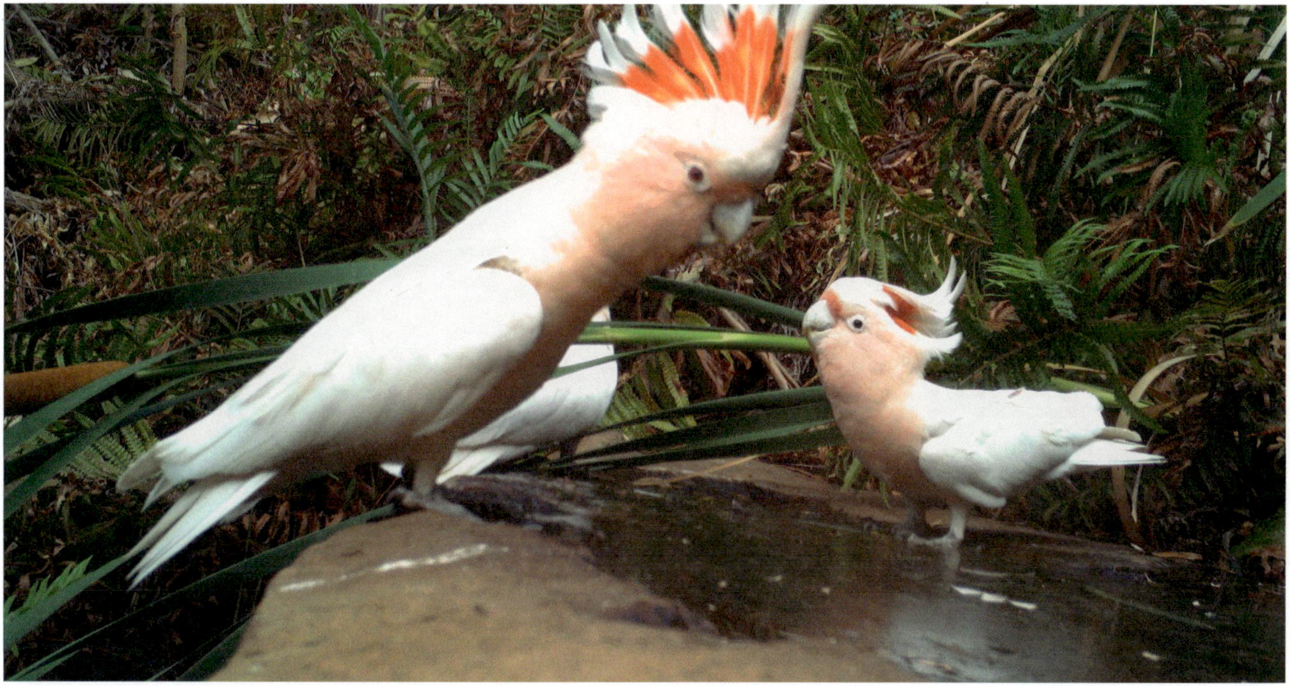

It was hard not to smile when this pair of Major Mitchell's cockatoos/kakalyalya's arrived at the permanent spring. Presumably they were just passing through the area and stopped for a drink.

Budgerigar/Kilykintarri, *Melopsittacus undulatus*

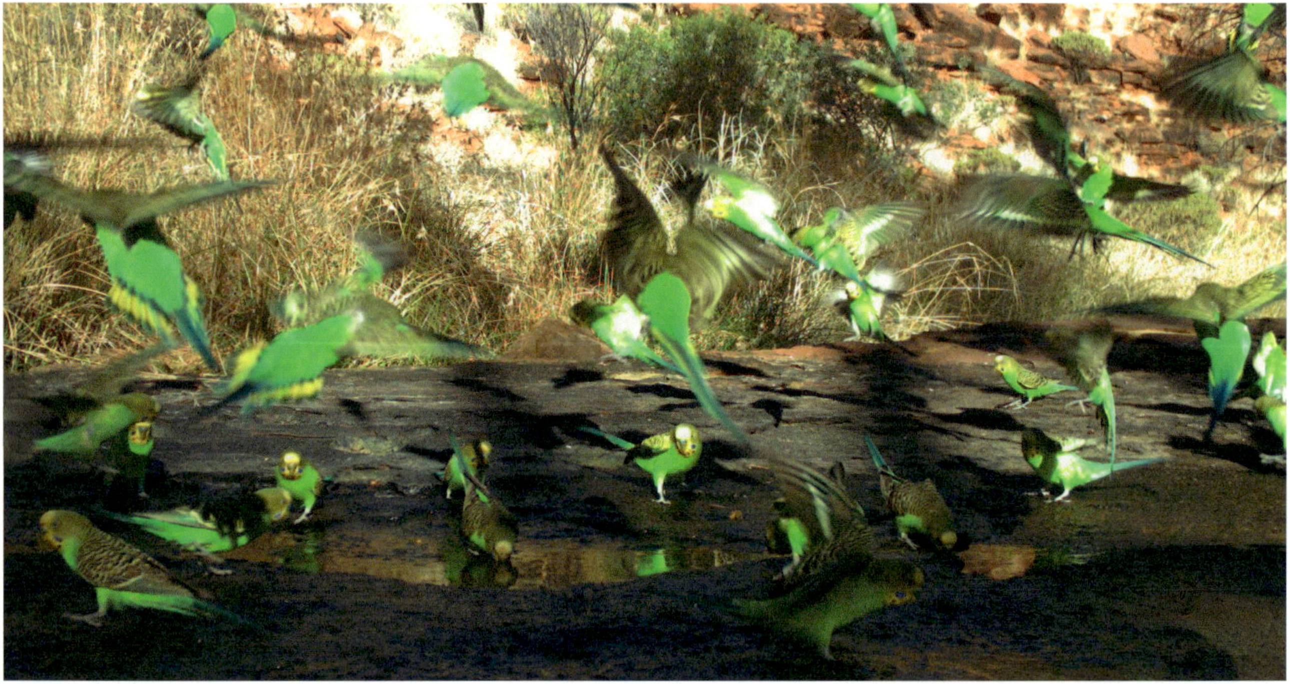

Budgerigars/kilykintarri, are a nomadic species that often occur in large flocks in open country after good rains have fallen.

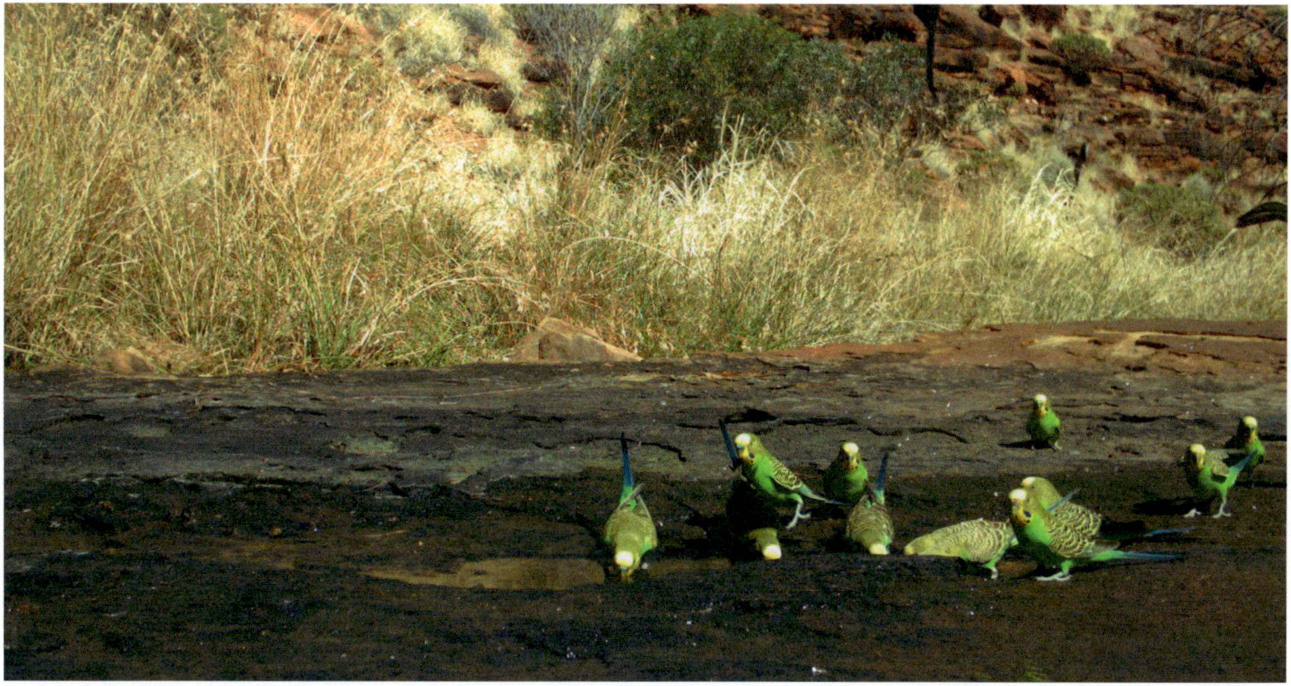

Two hours after their first arrival at the pool some budgies were still making the most of this water source.

Honeyeaters

Honeyeaters (nectarivores) are frequent visitors to Park waterholes. Most species can obtain water from their nectar-rich diets, but many supplement their diets with insects. Honeyeaters regularly appear at waterholes to drink, to bathe and to access the nectar and insects on nearby plants. They are small to medium-sized birds and different species can often be easily recognised, usually by a distinctively coloured stripe or feather pattern, or beak colour. White-plumed honeyeaters and grey-headed honeyeaters are the most commonly recorded small-bodied species at the waterholes. Larger-bodied species that regularly occur at the waterholes include spiny-cheeked honeyeaters and yellow-throated miners. Other species that you might see include the black-chinned honeyeater, the brown honeyeater, the grey-fronted honeyeater, the pied honeyeater, the singing honeyeater and the white-fronted honeyeater.

White-Plumed Honeyeater/Pinpal-pinpal, *Ptilotula penicillata*

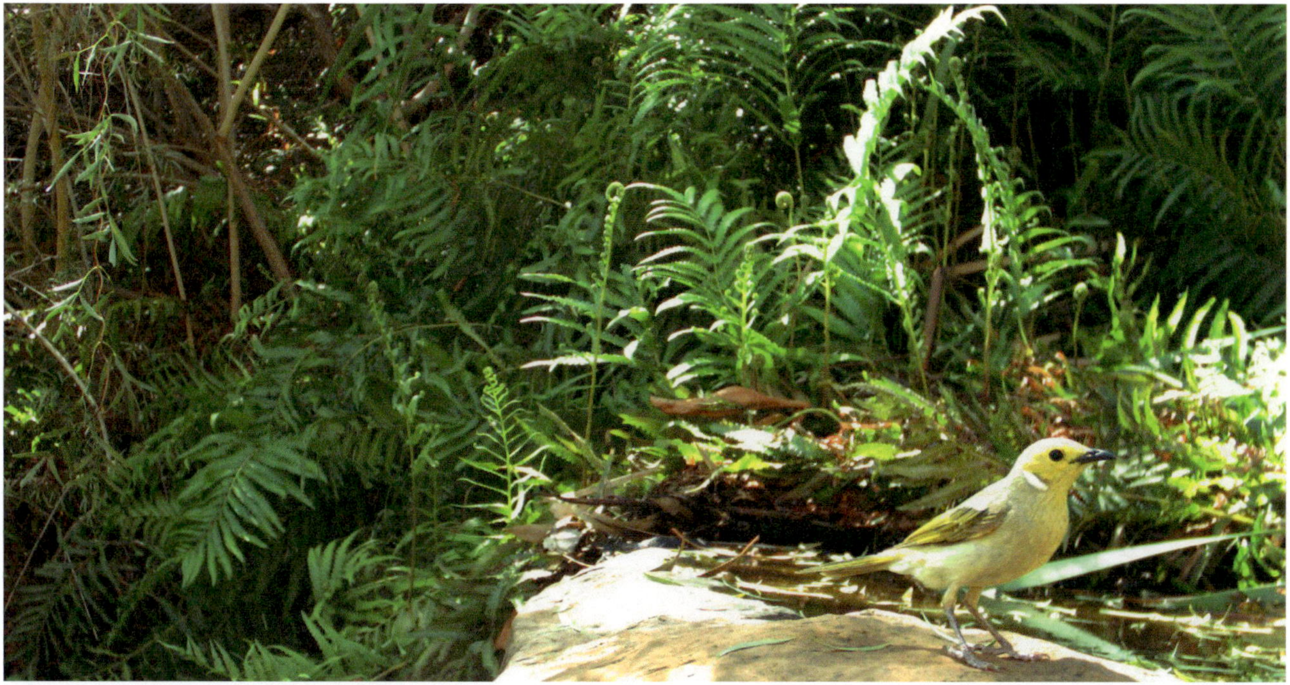

The white-plumed honeyeater / pinpal-pinpal, is the most commonly recorded honeyeater on the waterhole cameras. It often drinks with zebra finches and diamond doves.

Spiny-Cheeked Honeyeater/Titirara, and Pied Honeyeater, *Acanthagenys rufogularis* and *Certhionyx variegatus*

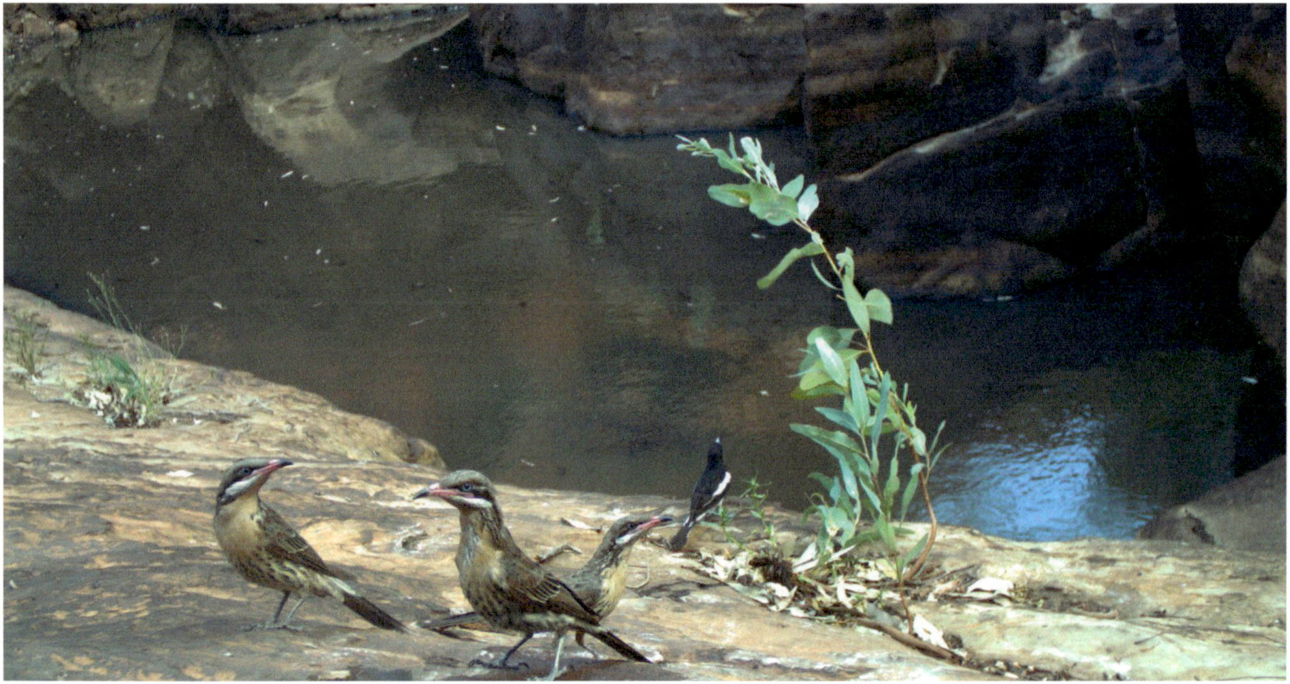

Spiny-cheeked honeyeaters/titirara appear to be guarding this groundwater seep. A pied honeyeater is visible in the background at the edge of the rockhole.

Yellow-Throated Miner/Piil-piilpa, *Manorina flavigula*

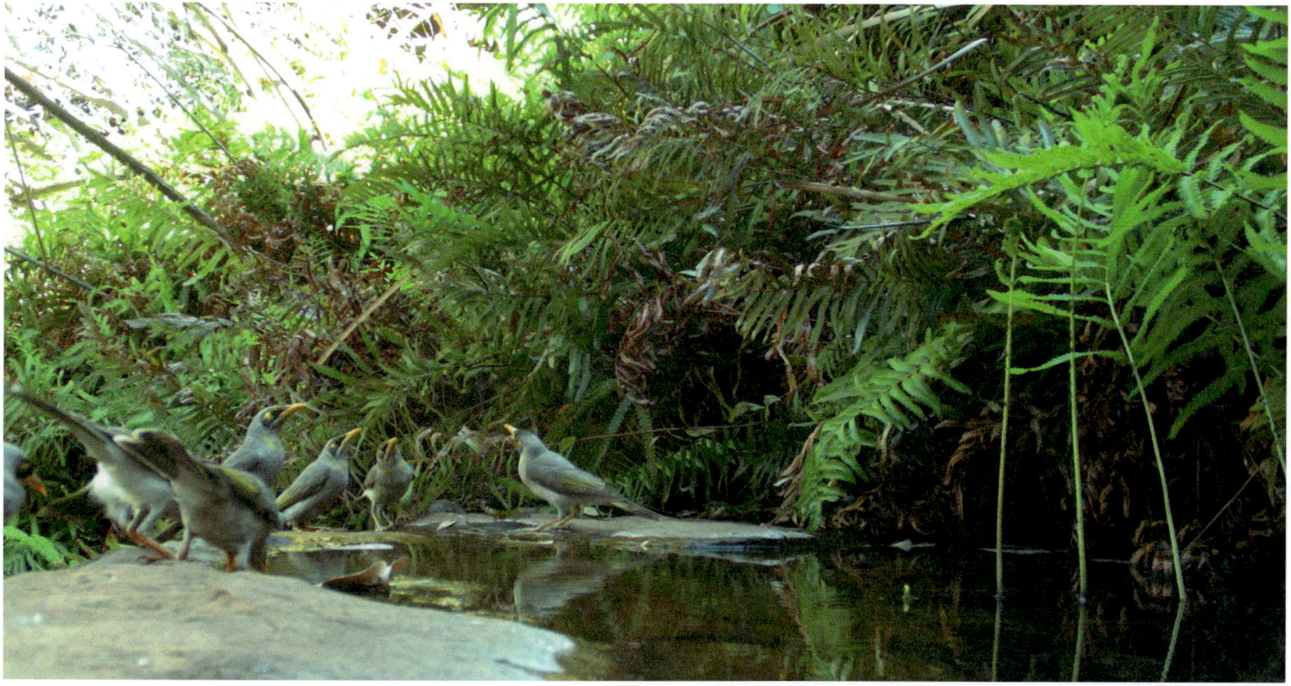

Yellow-throated miners/piil-piilpa, usually arrive at a waterhole in a group. They are an aggressive species and they rarely share a waterhole with other birds.

Waterbirds

It may seem a little odd to find birds that are dependent on water for most of their life cycle in inland arid Australia, where the only perennial water bodies are usually small springs, and the larger lakes are often dry salt pans or rarely hold water. Some of the waterholes in the Finke River, which arises in Tjoritja / West MacDonnell National Park and flows through the Finke Gorge National Park, contain fish, and these are an important source of food for visiting waterbirds. However, fish are absent from the isolated stream networks of Watarrka National Park. It is the resident frog populations, and large aquatic invertebrates, such as dragonflies, aquatic beetles and water scorpions, rather than fish, that sustain the waterbirds recorded on the waterhole cameras.

In addition to the species pictured overleaf, other waterbirds that you might see at the Watarrka waterholes include the Australasian grebe, grey teal, hoary-headed grebe, little black cormorant, and the Pacific black duck.

White-Faced Heron, *Egretta novaehollandiae*

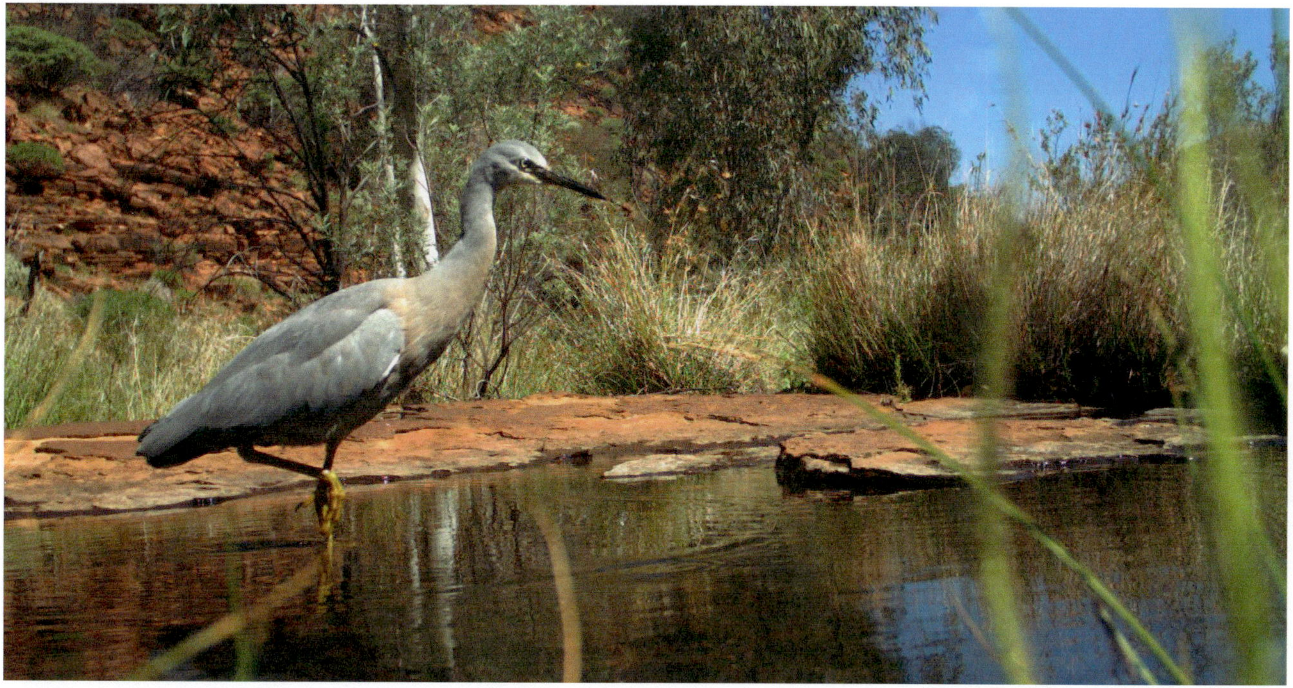

A white-faced heron hunting tadpoles, frogs and large aquatic invertebrates at the intermittent stream pool.

White-Necked Heron, *Ardea pacifica*

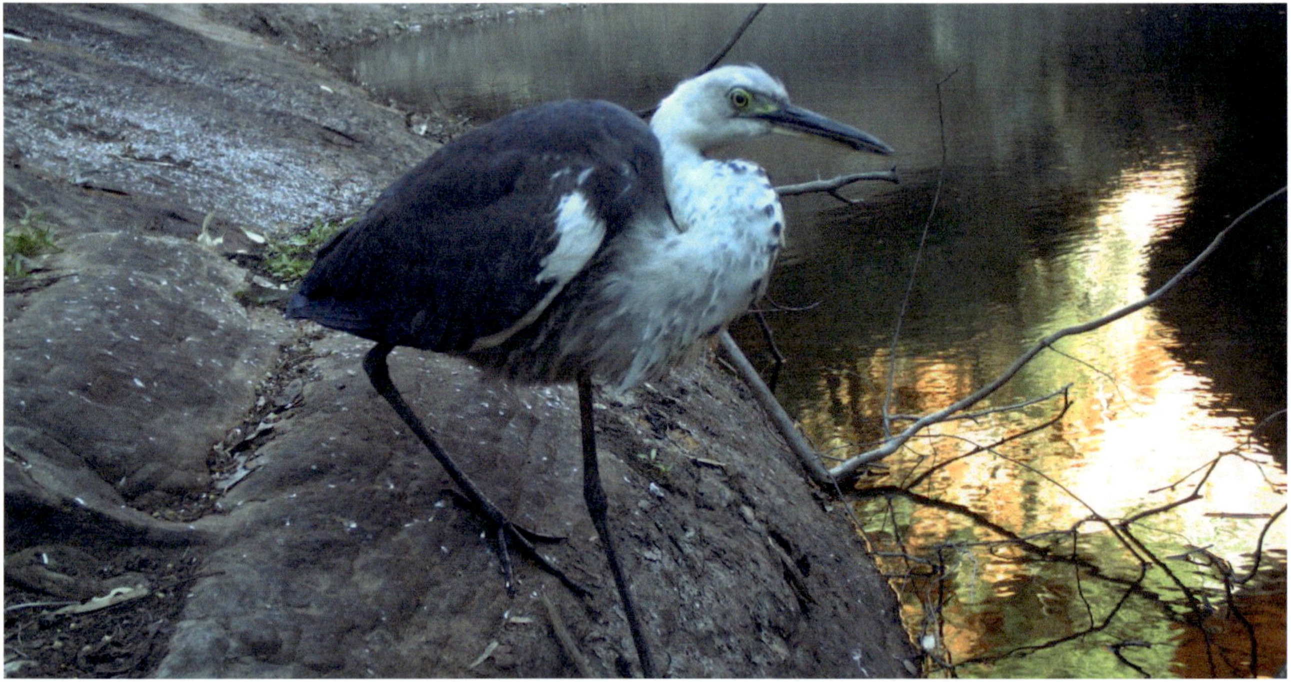

The white-necked heron, also known as the Pacific heron, has a distinctive double line of black spots running down the front of its neck and chest. This bird is fluffed up after swimming in the permanent rockhole.

Nankeen Night Heron, *Nycticorax caledonicus*

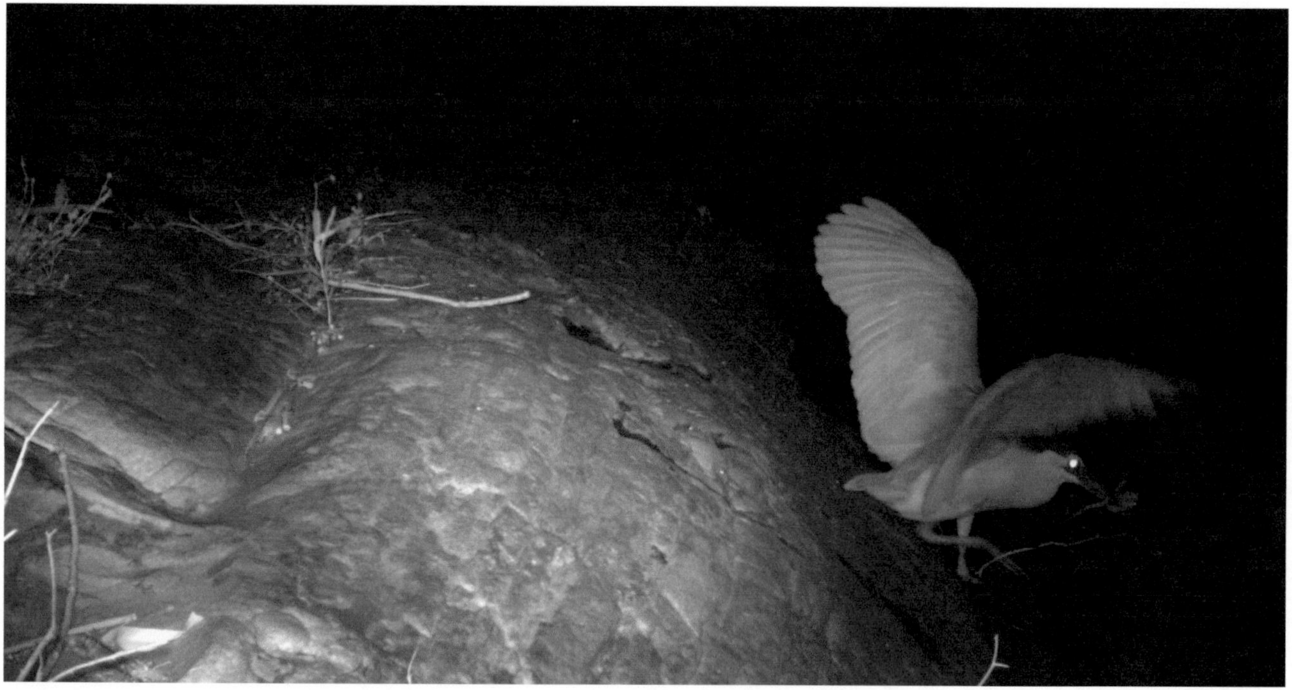

Photographic proof that the nankeen night heron preys on frogs at this rockhole.

Reptiles

The large reptiles that occur in the Park, including the perentie, *Varanus giganteus*, the black-headed monitor, *Varanus tristis*, and the Centralian carpet python, *Morelia bredli*, do not depend on waterholes to provide drinking water in the way that many birds do. Instead, they visit mostly to hunt frogs and other vertebrates. The perentie is Australia's largest lizard, it is endemic and occurs in arid regions across Queensland, NT, northern SA and Western Australia. The black-headed monitor is a smaller lizard and widespread across mainland Australia. Both are carnivores, preying on other reptiles, amphibians, small mammals, birds and insects.

The Centralian carpet python only occurs in the rocky range country of central Australia. In December 2019, the perennial spring camera recorded a Centralian carpet python striking a collared sparrowhawk. The python had been lying in wait within the pool, possibly for days, for its prey. It seems that a fierce battle was fought but the carpet python ultimately won.

Smaller reptiles are often seen near the waterholes, but are rarely captured on the cameras. Long-nosed dragons can often be seen in the sandy creek beds, especially during the warmer summer months. Central netted dragons can be encountered, often basking in the sun, on rocky outcrops throughout the Range.

The highly venomous mulga snakes, aka king brown snakes, are very occasionally recorded by the camera traps.Their images have only been recorded at night and they are probably seeking food, especially frogs, which are usually plentiful after rain has filled temporary pools.

Perentie, *Varanus giganteus*

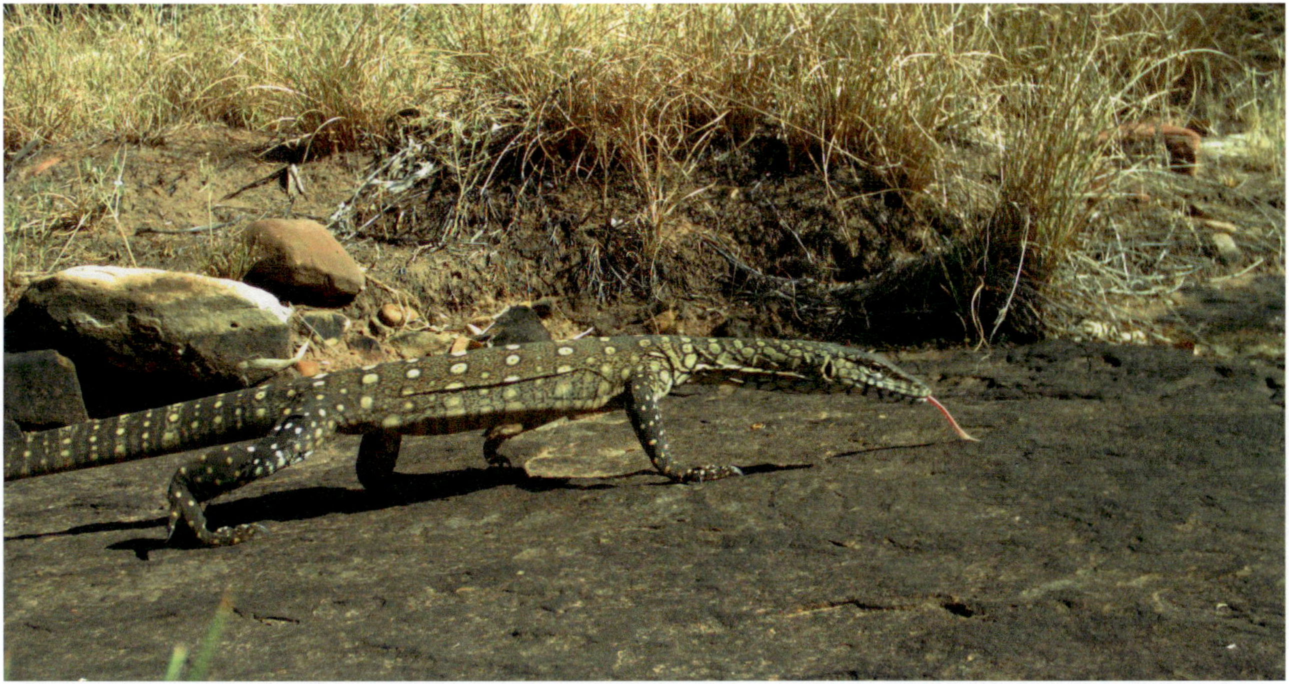

Perenties are only active during the warmer months of the year. Dry streambeds provide an easy pathway for these large reptiles to move through the Park.

Black-Headed Monitor, *Varanus tristis*

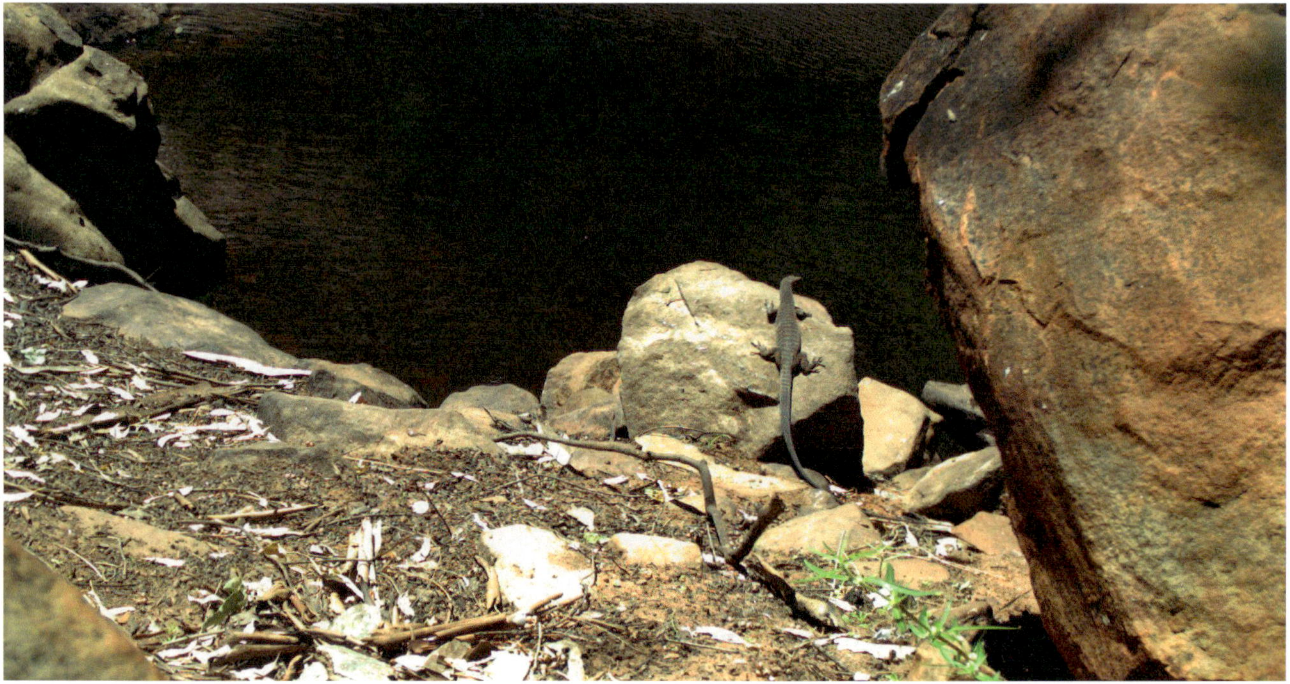

Black-headed monitors often appear during warmer weather. They have been recorded many times basking on the warm rocks at this permanent rockhole.

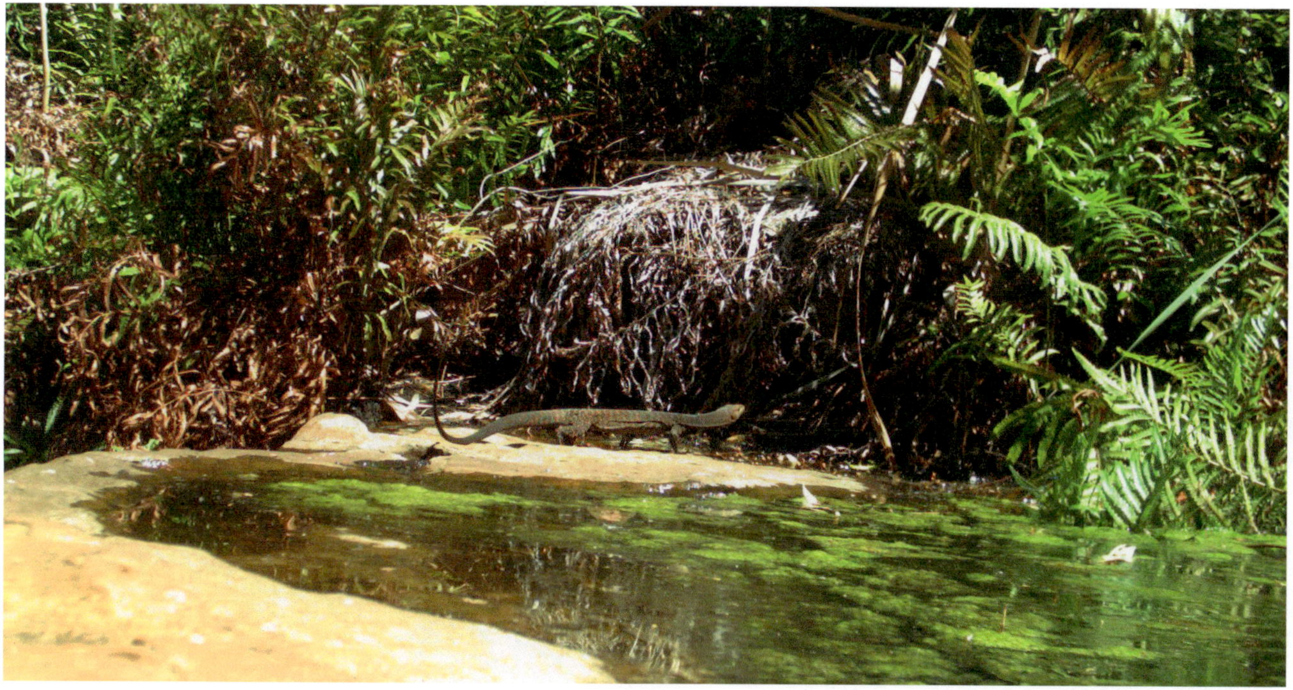

The dense bed of ferns sustained by constantly flowing groundwater at the perennial spring contains a resident population of the Centralian tree frog.The frogs shelter within the ferns during the day but a monitor on the hunt may still be able to catch one.

Centralian Carpet Python/Tatjalypa, *Morelia bredli*

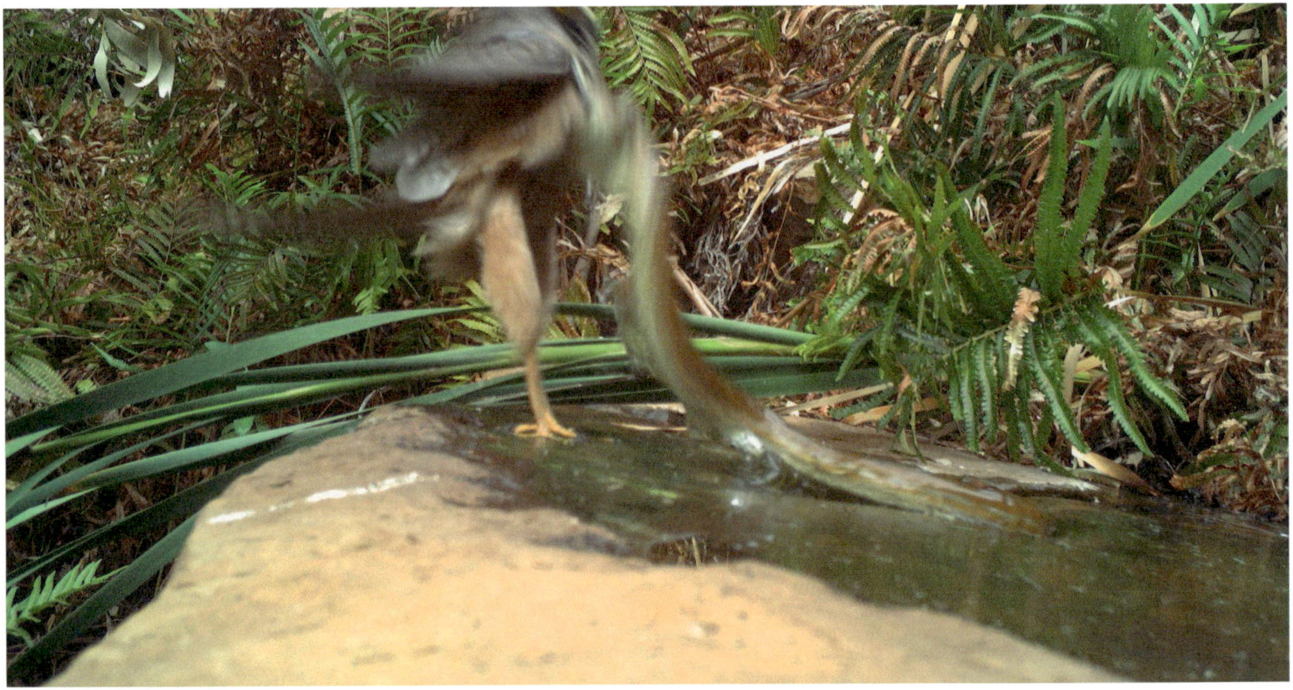

Life and death at the spring: this image records the moment that a Centralian carpet python/tatjalypa, rears up out of the water to take down the collared sparrowhawk that had come to drink.

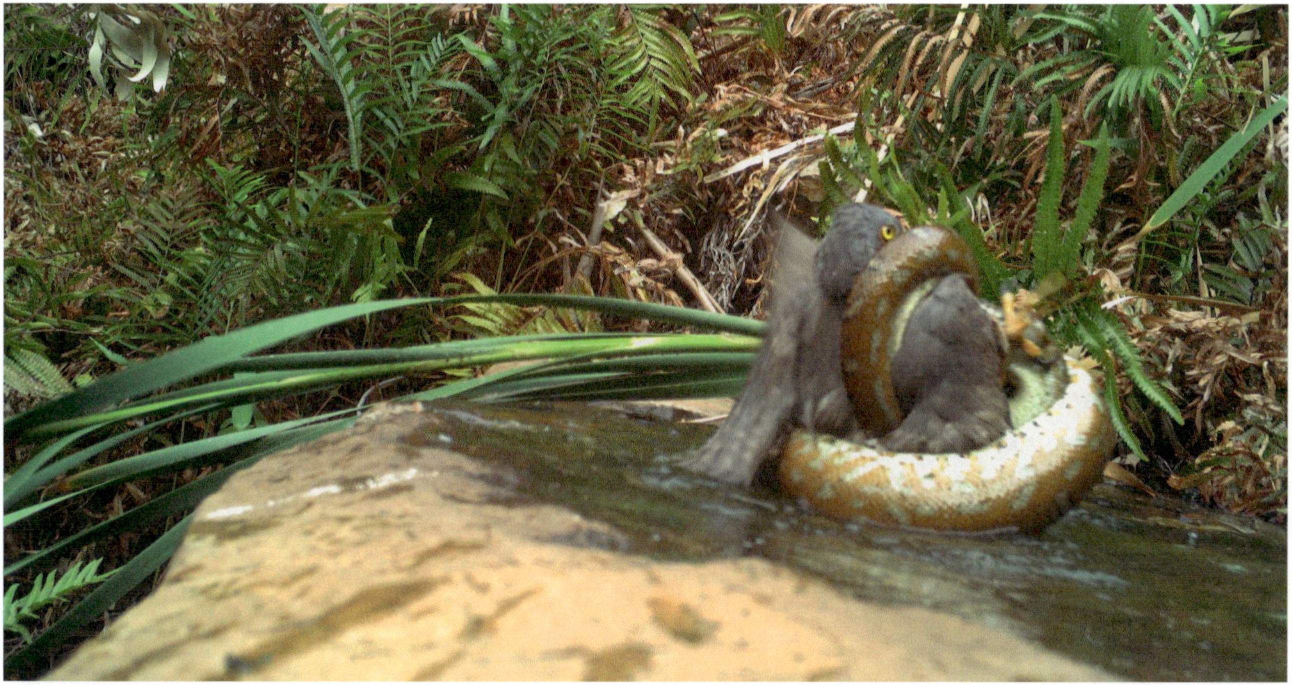

Although the collared sparrowhawk put up a fight, the Centralian carpet python ultimately won. Almost an hour later, the camera recorded a fatter python moving up the bank away from the pool.

Frogs

The two species most likely to be seen at Watarrka waterholes are the Centralian tree frog, *Litoria gilleni,* and the desert tree frog, *Litoria rubella.* The Centralian tree frog is the larger of the two species and closely resembles the common Australian green tree frog, *Litoria caerulea,* in colour and size. The desert tree frog is smaller and red-brown in colour. The former only occurs in central Australia whilst the latter is more widely distributed across the continent. Despite their names, neither species is closely associated with trees, instead both shelter in rocky crevices and within fern banks during the day and move onto the edges of waterholes at night. Frog populations boom after big rain events but small populations persist at Park waterholes, even during the driest of times.

If you do visit the Park soon after heavy rain, you may see the burrowing frog, also known as the water-holding frog,*Cyclorana platycephala.* These stout and distinctively patterned frogs spend most of their life underground, cocooned in burrows within moist sandy streambeds. They emerge after rains to breed and lay eggs in temporary stream pools. So far these frogs have not been recorded on the waterhole cameras, probably because their time on the surface is very short-lived.

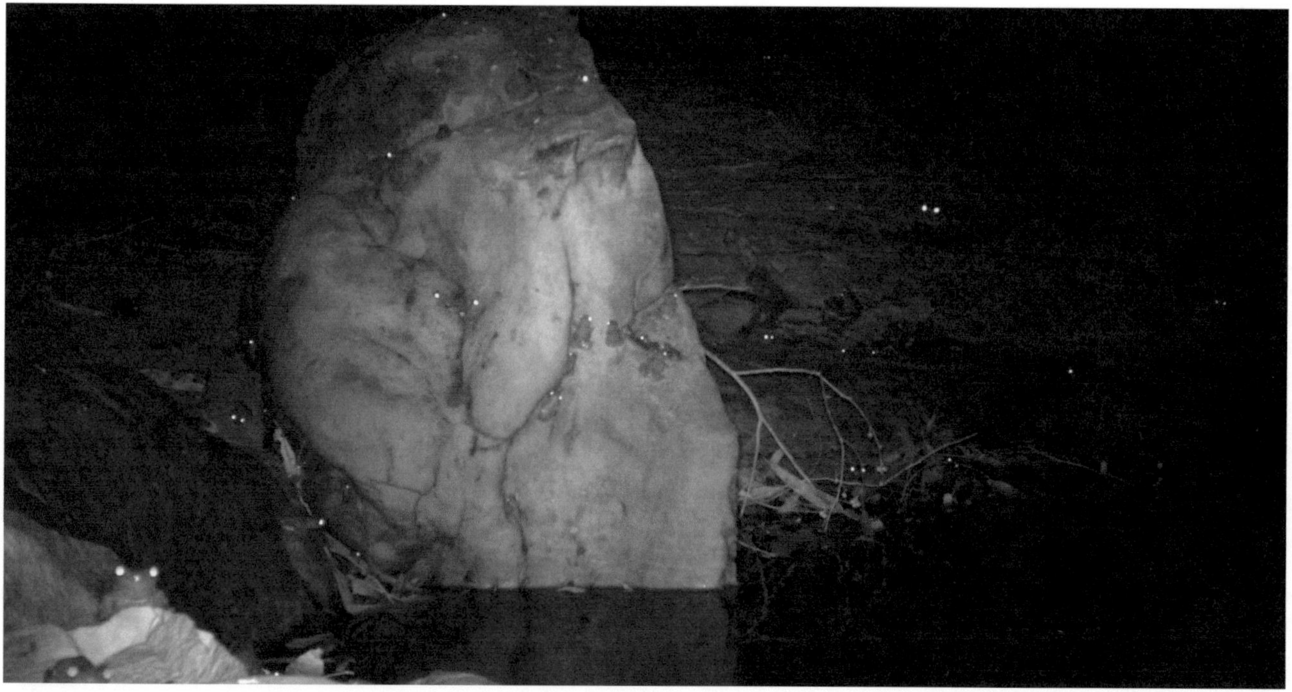

Their eye shine provides a clue as to the presence and numbers of frogs at this waterhole.

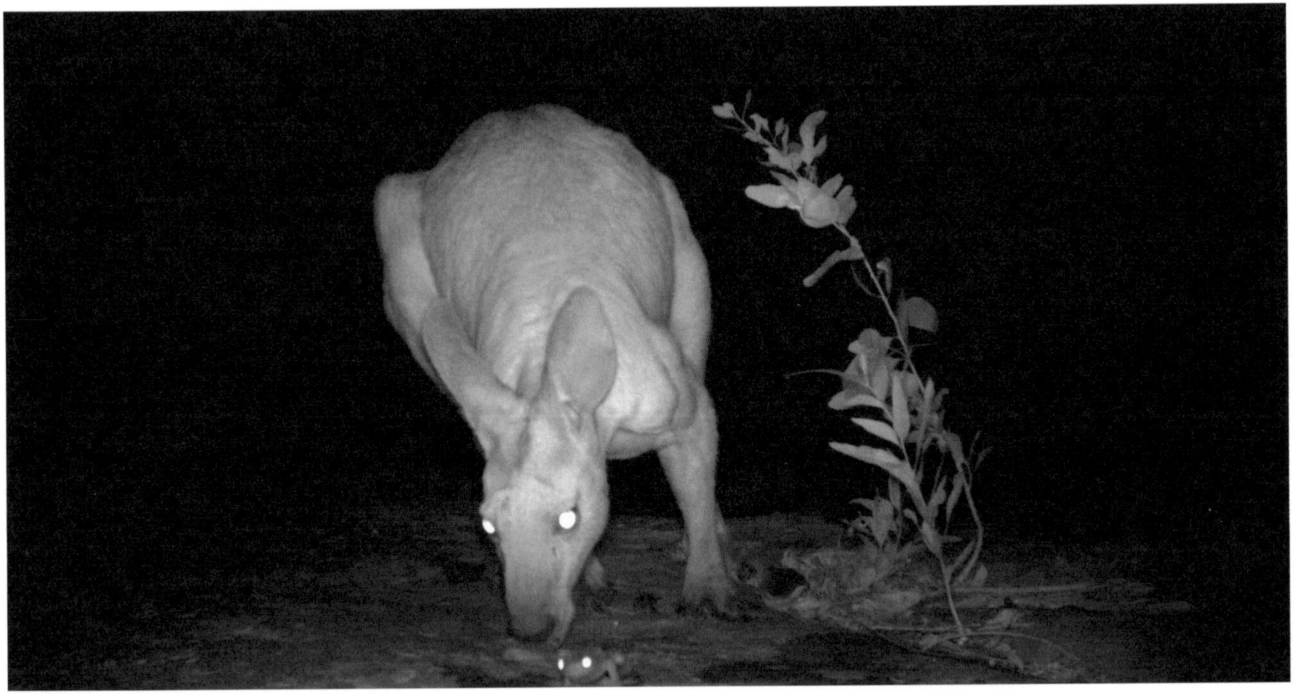

Waterhole frogs rarely trigger a camera, but they often photobomb euro and dingo images.

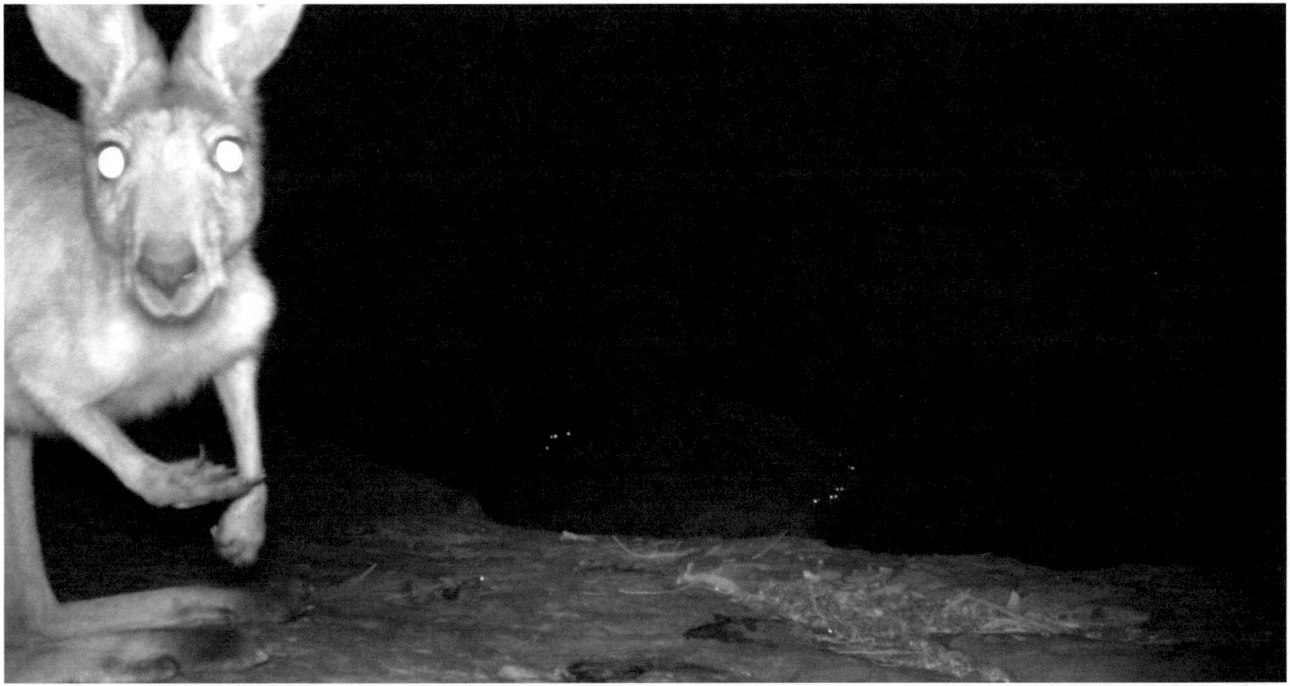

Those frogs are with me…

Feral Animals

The importance of Watarrka National Park for the conservation of biodiversity in central Australia is highlighted by the low number of feral animals captured on camera, despite the hundreds of thousands of images that have been recorded to date. Occasionally a feral cat, *Felis catus*, will appear at a waterhole, but cat densities appear to be low. The introduced red fox, *Vulpes vulpes*, occurs in the Park, but is rarely recorded on the waterhole cameras, apparently preferring the drier rocky outcrops and adjacent sandy plains. Ideally, neither feral cats nor red foxes should be present in the Park and new approaches to the control of these species may mean that they will be eliminated in the future.

Although camels and horses are present on lands close to the Park boundaries, neither species has been recorded on the waterhole cameras. Feral camels and feral horses degrade water bodies through drinking, trampling and fouling with dung. The negative impacts of feral camels on central Australian waterholes are now well known. They have been described in papers by Brim Box *et al.* (2010, 2016) and an experimental study by McBurnie *et al.* (2014) found that camel dung reduces water quality and results in a decrease in the number and types of animals that can use a water source.

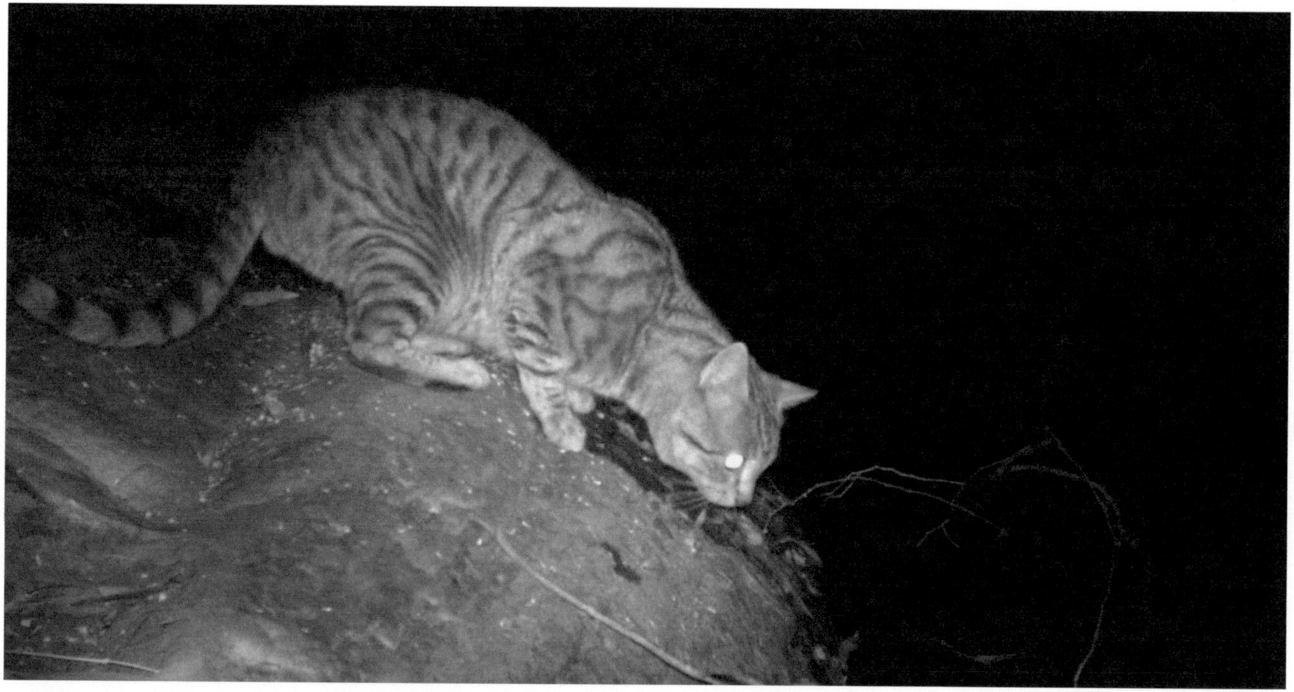

A feral cat at the permanent rockhole.

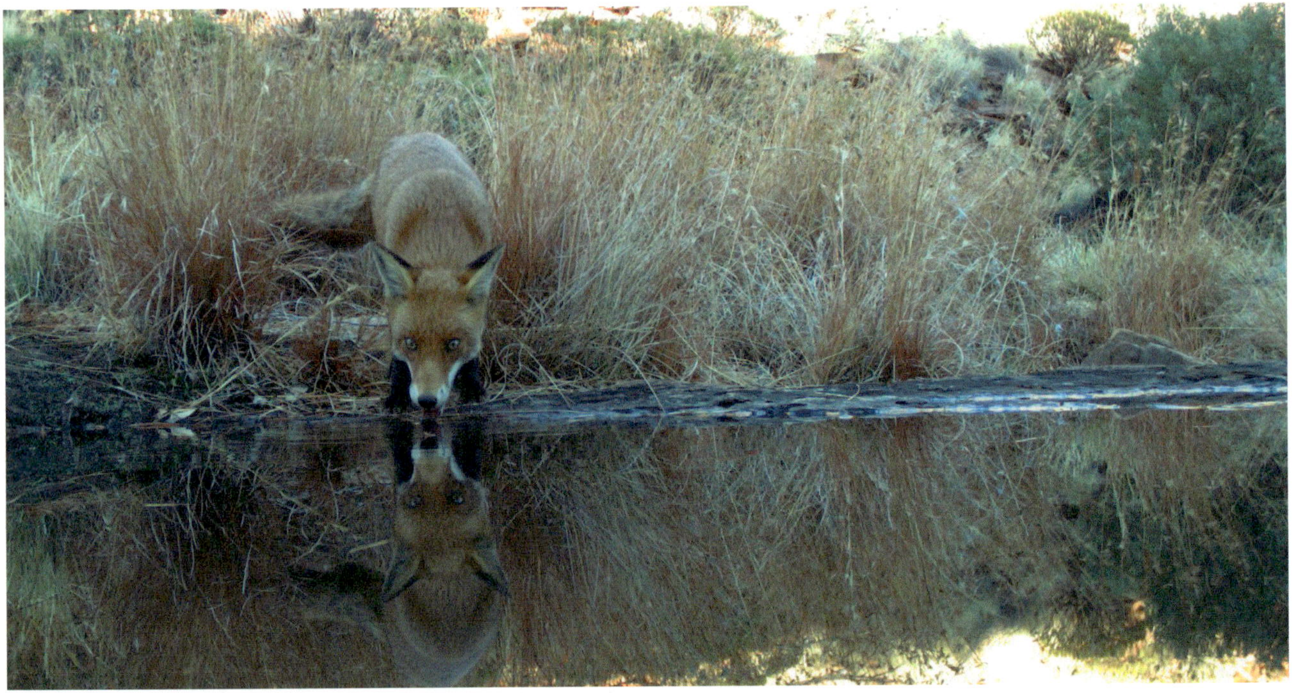

A red fox and its reflection: a beautiful image of an animal that should not be here.

Camera Trap Challenges

Camera trap images have not only revealed much about the species that come to the Watarrka waterholes to drink, they also provide valuable hydrological and ecological information. This includes visual records of when the waterholes contain water, or are dry, and the condition of the surrounding vegetation. The remoteness of the Park means that this hydrological and ecological information is not currently recorded in any other way. The long periods between rainfall events in central Australia means that cameras can be placed close to the edge of a spring or waterhole with very little risk of being flooded. However, when rain does fall, it is often a very large event and flash flooding can occur. Some cameras have been lost in these flash floods, but overall the loss of cameras in this project has been very low.

Interpreting camera trap images can be both a science and an art. Sometimes the identification of a species is based on the minimum of photographic information. However, even a small part of an animal captured on camera can be enough for it to be identified, and thereby its presence recorded.

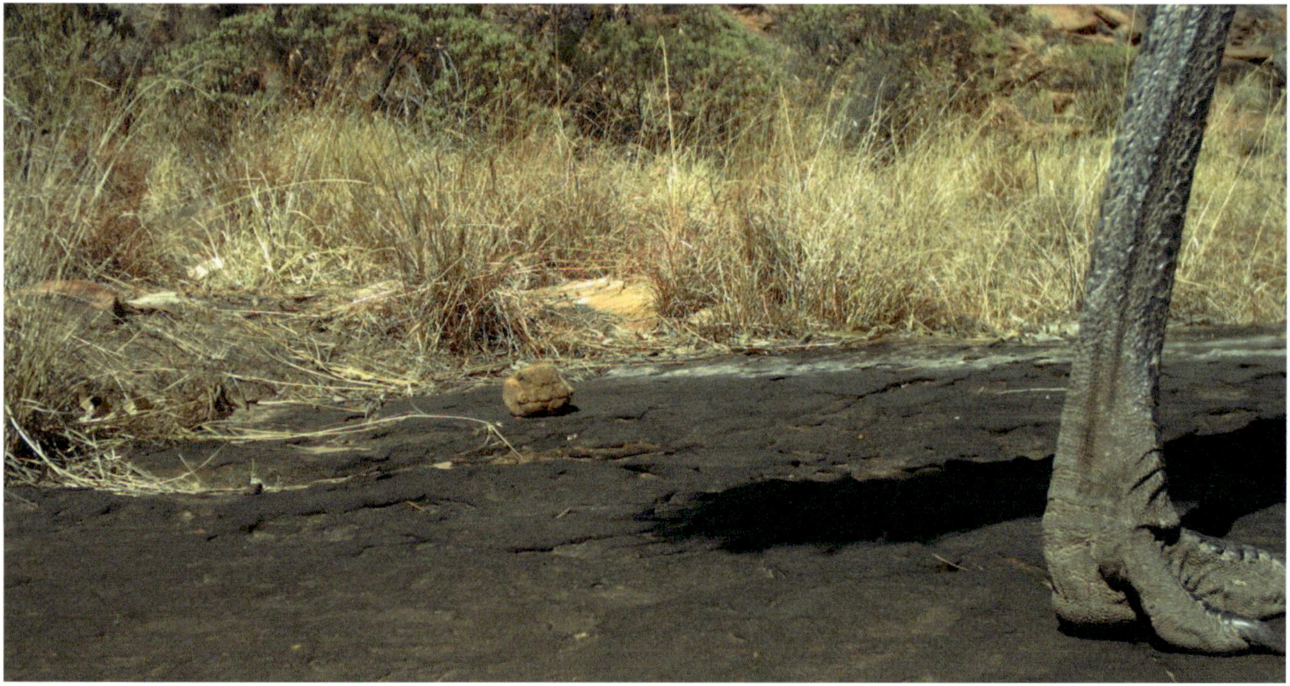

Although only part of this species is visible, it is not difficult to conclude that this leg belongs to an emu/kalaya, *Dromaius novaehollandiae*.

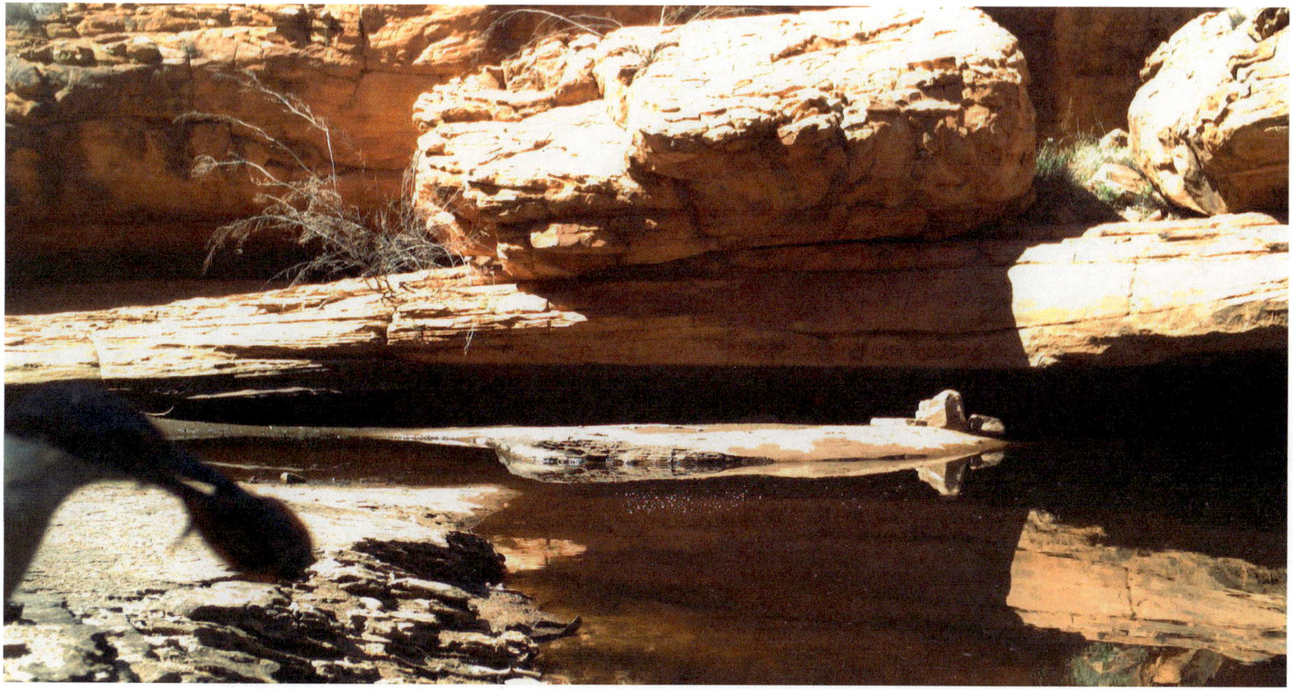

This image is very informative, despite the poor focus. A sacred kingfisher/luunpa, *Todiramphus sanctus*, is holding its prey, a shield shrimp, *Triops australiensis*. Shield shrimps only appear when temporary arid zone pools fill with rain. They survive long dry periods in the sediments as resistant eggs.When the pool fills, the eggs hatch and the shrimps grow quickly to complete their life cycle before the pool dries again,usually within three to four weeks.

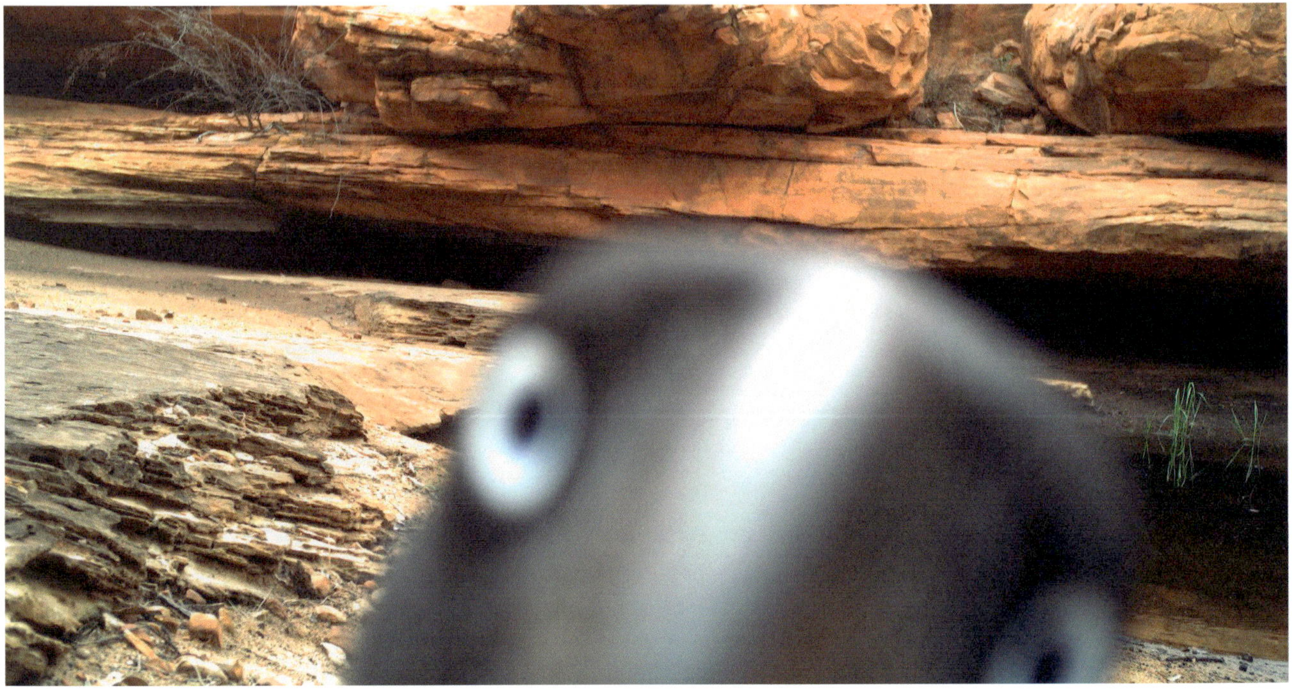

A curious crow proves that an image does not have to be in focus to make you laugh.

Conservation Status of the Park and its Wildlife

Watarrka National Park is classified as an IUCN Category II protected area. This means that tourism and local business enterprises that operate within the Park need to be compatible with conservation objectives. The remoteness of the Park contributes to the conservation of its wildlife because it is not subject to the pressures that often accompany densely settled areas and large cities. However, its remoteness is also a vulnerability. Species that are lost from the Park may not be able to recolonise because the large tracts of arid land that surround it are a major barrier to species dispersal.

The IUCN Red List Categories and Criteria (iucnredlist.org) provides a universally understood and accepted system for classifying species at high risk of global extinction. Species are divided into nine categories: Not Evaluated, Data Deficient, Least Concern, Near Threatened, Vulnerable, Endangered, Critically Endangered, Extinct in the Wild and Extinct. All the avian species recorded at the Watarrka waterholes are of least concern (LC) on the IUCN Red List. Of the mammals and reptiles recorded by the waterhole cameras so far, it is only the warru, the black-footed rock-wallaby, that is considered vulnerable (VU).

A long term study of aquatic biodiversity at sites in Watarrka and Tjoritja/West MacDonnell National Parks has revealed the critical role that the Parks play in protecting relictual aquatic species, particularly through the exclusion of feral animals, especially camels (Brim-Box *et al.* 2014). However, we can never be complacent as species that are endemic (restricted in their distribution) to central Australia, and those at the top of food chains, are especially vulnerable to environmental change. A species that is common can become rare very quickly through changes to its environment brought about by climate change, invasive and feral species, habitat destruction, water extraction and other human impacts.

The Importance of Groundwater

Groundwater is the critical resource that supports much of the wildlife in central Australia (Box *et al.* 2008, Davis *et al.* 2020). Elsewhere in Australia, the amount of groundwater available to the environment is declining through over-extraction, while water quality is degraded by pollution from urban, industrial, mining and agricultural activities. The remoteness of the Park, particularly its distance from the large cities and towns of Australia's eastern coast, has provided protection from the impacts of pumping groundwater to supply urban populations. The region's aridity, combined with infertile soils, means that neither intensive agricultural enterprises, nor horticultural developments, have occurred at the large scales that have had a major impact on water quantity and quality elsewhere.

The potential adverse impacts from gas extraction, known as fracking, raised concerns when Palatine Energy applied for permission to explore for shale gas and oil across the Park in 2012. The company indicated that it would use hydraulic fracturing (fracking) to extract tight oil. After repeated requests from the Traditional Owners ('TOs') to the NT government to protect the Park from fracking, a group of TOs, supported by the Central Land Council and the NT Environmental Defenders Office went to Canberra in November 2015, to meet with the Federal Environment Minister, Greg Hunt. They were seeking emergency protection under federal heritage laws after Territory governments had ignored their appeals. The TOs' concerns included the over-use and contamination of the Park's surface and groundwater resources that support much of the bush tucker that they traditionally hunt and gather. Finally, in May 2017, the NT government declared a mineral and petroleum reserve over the entire park. This declaration, under the Petroleum Act, provides a significant level of protection to the Park that excludes any future mineral or energy exploration and extraction.

From Research Datasets to Park Management

The waterholes of the Park, and other protected areas in the central Australian ranges, including the Tjoritja/West MacDonnell and Finke Gorge National Parks, provide unique opportunities for research on systems that are largely unaffected by the impacts that prevail in more densely populated regions. The ongoing operation of the cameras provides the long-term datasets that are essential for understanding changes in species and habitats over time. This is particularly important in a highly variable climatic region such as central Australia.

Waterhole camera datasets have formed the basis of an Honours and a PhD project undertaken by Simon Votto at the University of Canberra and Charles Darwin University. Some of Simon's work on the importance of the waterholes for terrestrial birds during hot, dry periods, has already been published in scientific journals (Votto *et al.* 2020 & 2022) and further publications are planned. The camera datasets were also integral to a study showing how environmental DNA metabarcoding can be used to detect vertebrates at arid zone waterholes (Furlan *et al.* 2020).

More generally, research undertaken at Watarrka waterholes has contributed to scientific papers that describe the importance of groundwater-dependent springs as current and future refugia for both aquatic and terrestrial species (Davis *et al.* 2013, 2017). How rainfall events, landscape features and species mobility influence the dispersal of species across large tracts of arid land was the focus of Emma Razeng's PhD study (Razeng *et al.* 2016, 2017). The first description of the microbial assemblages in waterbodies at Watarrka and other central Australian national parks was provided by Mirjam Kaestli (Kaestli *et al.*, 2019). Citizen scientists have also been involved in research at Park waterholes through the collection of water samples to identify groundwater-fed refugia (Davis *et al.* 2020).

The waterhole camera trapping project has been particularly valuable in recording avian biodiversity. A total of 70 species of birds have been recorded by the waterhole cameras since 2014. This represents more than half the total number (126) of avian species previously recorded in surveys across multiple Park habitats. Not surprisingly, fewer records of other less water-dependent groups have been obtained, including 7/20 mammal species, 4/71 reptiles and 2/7 frog species. The success of this small spatial study, undertaken at only four waterholes, shows that camera traps are an important tool for Park biodiversity monitoring and management.

Although the cameras have provided an almost continuous record of the presence of vertebrate species at waterholes since 2014, some technological challenges still remain. The location of the waterholes within shaded gullies means that the use of solar panels to power the cameras is logistically difficult. Current operations require replacing the rechargeable camera batteries and SD cards at regular intervals. The lack of cheap and easy access to the internet in this remote region means that live streaming of images is not yet possible. However, the rapid pace of technological development may mean that a Watarrka waterhole webcam could be established fairly soon.

This study started as a pilot project to determine whether it is possible to collect wildlife data from remote waterhole sites on a continuous basis.The answer to that research question is an emphatic yes. Some of the anticipated problems, for example, cameras being displaced or washed away by flash floods, have rarely occurred.The cameras have proved to be extremely robust and camera trap technology is likely to continue to improve. New methods and technologies will speed up camera trap data analysis and interpretation. For example, machine learning will facilitate the identification of species, especially species of management concern, such as rare species and feral species. It will also decrease the time spent processing multiple repeat images such as those triggered by moving (wind-blown) vegetation.

The longer this project continues, the more valuable the information it generates becomes. Long-term datasets are vitally important for the development and assessment of evidence-based climate adaptation and conservation programs.Waterhole camera trap data can serve the dual purpose of supporting local wildlife management whilst also contributing to the global knowledge and understanding of arid zone aquatic ecosystems.

References

Ara Winki - Stories from Aṟa Irititja - AppBooks

Atlas of Living Australia - www.ala.org.au.

Aumann, T (1990). Use of stones by the Black-breasted Buzzard *Hamirostra melanosternon* to gain access to egg contents for food. *Emu* 90: 141–144.

Brim-Box J, Davis J, Strehlow K, McBurnie G., Duguid A, Brock C, McConnell K, Day C, and Palmer, C. (2014). Persistence of central Australian aquatic invertebrate communities. *Marine and Freshwater Research* 65: 562–572.

Brim-Box, J, Guest, T, Barker, P, Jambrecina, M, Moran, S, Kulitja, R (2010) Camel usage and impacts at a permanent spring in central Australia: a case study.*The Rangelands Journal* 32: 55-62.

Brim-Box, J., G. McBurnie, K. Strehlow, T. Guest, M. Campbell, A. Bubb, *et al.* (2016). The impact of feral camels (*Camelus dromedarius*) on remote waterholes in central Australia. *The Rangeland Journal* 38:191-200.

Box, JB, Duguid, A, Read, RE, Kimber RG, Knapton, A, Davis JA &. Bowland, AE (2008) Central Australian waterbodies: the importance of permanence in a desert landscape. *Journal of Arid Environments* 72: 1395-141

Davis, J.A. (1986). Revision of the Australian Psephenidae (Coleoptera): Systematics, Phylogeny and Historical Biogeography. *Australian Journal of Zoology*, 119:1-97.

Davis, JA, Munksgaard, N, Hodgetts,J & Lambrinidis, D (2020) Identifying groundwater-fed climate refugia in remote arid regions with citizen science and isotope hydrology *Freshwater Biology* DOI: 10.1111/fwb.13601

Davis JA, Sim, L, Sunnucks, P, Thompson, RM, Pinder, A, Brim-Box, J, Murphy, NP, Sheldon, F, Morán-

Ordóñez, A & Sunnucks, P (2018) Patterns and drivers of aquatic invertebrate diversity across an arid biome *Ecography* 41: 375–387.

Davis, JA, Kerezy, A & Nicol, S (2017) Conserving perennial water is critical in arid landscapes. *Biological Conservation* 211: 30–35.

Davis, JA, Pavlova, A, Thompson, R & Sunnucks, P (2013) Evolutionary refugia and ecological refuges: key concepts for conserving Australian arid zone freshwater biodiversity under climate change. *Global Change Biology* 19:1970–1984.

Davis, JA (1997) Conservation of aquatic invertebrate communities in central Australia. *Memoirs of the Museum of Victoria* 56: 491-503.

Davis, JA (1996) Aquatic ecosystems in central Australia: comparison of recent records of fish and invertebrates with those of the Horn Expedition. Exploring Central Australia: Society, the Environment and the 1894 Horn Expedition, edited by S. R Morton and D. J. Mulvaney. Surrey Beatty and Sons, Chipping Norton.

Duguid, A, Barnetson, J, Clifford, B, Pavey, CR, Albrecht, D, Risler, J & McNellie, M. (2005) Wetlands in the arid Northern Territory. A report to the Australian Government Department of the Environment and Heritage on the inventory and significance of wetlands in the arid NT. Alice Springs, Northern Territory Government Department of Natural Resources, Environment and the Arts.

Field *et al.* (2022) The Australian dingo is an early offshoot of modern breed dogs
Science Advances 8: 1-14, eabm5944.

Furlan, EM, Davis, JA & Duncan, RP (2020) Identifying error and accurately interpreting eDNA metabarcoding results: A case study to detect vertebrates at arid zone waterholes *Molecular Ecology Resources* doi.org/10.1111/1755-0998.13170

Kaestli M, Munksgaard N, Gibb K, Davis J (2019) Microbial diversity and distribution differ between water column and biofilm assemblages in arid-land waterbodies. *Freshwater Science* 38: 869-882.
Giles, E (1889). Australia Twice Traversed. Facsimile edn. 1981. Doubleday, Sydney.
Gilfillan, S (2001a). An ecological study of a population of *Pseudantechinus macdonnellensis* (Marsupialia: Dasyuridae) in central Australia. I. Invertebrate food supply, diet and reproductive strategy. *Wildlife Research* 28: 469–480.
Gilfillan, S (2001b). An ecological study of a population of *Pseudantechinus macdonnellensis* (Marsupialia: Dasyuridae) in central Australia. II. Population dynamics and movements. *Wildlife Research* 28: 481–492.
Hansen KC & Hansen, LE (1992) Pintupi/Luritja Dictionary 3rd Edition, Institute for Aboriginal Development, Alice Springs.
Latz, PK, Johnson, KA & Gillam, MW (1981) A biological survey of the Kings Canyon area of the George Gill Range. Internal Report . Conservation Commission on the NT. Botanical data updated 1985.
Lavery, TH, Eldridge, M, Legge, S *et al.* (2021)Threats to Australia's rock-wallabies (*Petrogale* spp.) with key directions for effective monitoring. *Biodiversity and Conservation* 30: 4137–4161.
Letnic, M, Koch, F, Gordon, C, Crowther, MS, Dickman, CR (2009) Keystone effects of an alien top-predator stem extinctions of native mammals. *Proceedings of the Royal Society B* 276:3249-3256.
McBurnie, G, Davis, JA, Thompson, RM, Nano, C. & Brim-Box, J (2015) Experimental evidence for impacts of an invasive herbivore (*Camelus dromedarius*) on arid zone freshwater pools. *Journal of Arid Environments* 113: 69-76.
Murphy, A, Pavlova, A, Thompson R, Davis, J, & Sunnucks, P, (2015) Swimming through sand: connectivity of aquatic fauna in deserts. *Ecology and Evolution* 5: 5252–5264.
Razeng, E, Smith, A, Harrisson, K, Pavlova, S, Nguyen, T, Pinder, A, Suter, P, Webb, J, Gardner, M, Brim

Box, J, Thompson, R, Davis, J, Sunnucks, P (2017) Evolutionary divergence in freshwater insects with contrasting dispersal capacity across a sea of desert. *Freshwater Biology* 62: 1443–1459.

Smith, BP, Cairns, KM, Adams, JW *et al.* (2019) Taxonomic status of the Australian dingo: the case for *Canis dingo*Meyer 1793. *Zootaxa* 4564:173-197.

Spencer, WB (1896) Report of the Horn Scientific Expedition to Central Australia. Melville,Muellen and Slade: Melbourne.

Votto, SE, Dyer, FJ, Caron, V & Davis, JA (2020) Thermally-driven thresholds in terrestrial avifauna waterhole visitation indicate vulnerability to a warming climate.*Journal of Arid Environments* 181:104217

Votto, SE, Schlesinger, C, Dyer, FJ, Caron, V & Davis, JA (2022) The role of fringing vegetation in supporting avian access to arid zone waterholes *Emu-Austral Ornithology* DOI: 10.1080/01584197.2022.2041441

Watarrka National Park Joint Management Plan (2010, 2018) Parks & Wildlife Commision.

List of all species recorded from 2014 — 2022

This section lists all species which were recorded by the waterhole cameras between 2014 and 2022.

Birds

- Australasian Grebe (*Teachybaptus novaehollandiae*)
- Australian Magpie (*Gymnorhina tibicen*)
- Australian Raven (*Corvus coronoides*)
- Australian Ringneck / Patilpa (*Barnardius zonarius*)
- Barn Owl (*Tyto javanica*)
- Black-Breasted Buzzard / Aralapalpalpa (*Hamirostra melanosternon*)
- Black Honeyeater (*Sugomel niger*)
- Black-Chinned Honeyeater (*Melithreptus gularis*)
- Black-faced Cuckoo-shrike (*Coracina novaehollandiae*)
- Brown Falcon / Kirkinpa (*Falco berigora*)
- Brown Goshawk / Kirkinpa (*Accipiter fasciatus*)
- Brown Honeyeater (*Lichmera indistincta*)
- Budgerigar / Kilykintarri (*Melopsittacus undulatus*)
- Cockatiel (*Nymphicus hollandicus*)
- Collared Sparrowhawk (*Accipiter cirrocephalus*)
- Common Bronzewing / Manpi (*Phaps chalcoptera*)

- Crested Bell Bird (*Oreoica gutturalis*)
- Crested Pigeon / Ngatapuka (*Ocyphaps lophotes*)
- Crimson Chat (*Epthianura tricolor*)
- Diamond Dove (*Geopelia cuneata*)
- Eastern Grass Owl (*Tyto longimembris*)
- Emu / Kalaya (*Dromaius novaehollandiae*)
- Fairy Martin (*Petrochelidon ariel*)
- Grey Butcherbird (*Cracticus torquatus*)
- Grey Shrike-Thrush (*Colluricincla harmonica*)
- Grey-Fronted Honeyeater (*Ptilotula plumula*)
- Grey-Headed Honeyeater (*Ptilotula keartlandi*)
- Grey Honeyeater (*Conopophila whitei*)
- Grey Teal (*Anas gracilis*)
- Grey Wagtail (*Motacilla cinerea*)
- Hoary-Headed Grebe (*Poliocephalus poliocephalus*)
- Horsfield's Bronze-Cuckoo / Nyiiru (*Chalcites basalis*)
- Little Black Cormorant (*Phalacrocorax sulcirostris*)
- Little Button Quail (*Turnix (Alphaturnia) velox*)
- Little Eagle (*Hieraaetus morphnoides*)
- Little Woodswallow (*Artamus minor*)
- Magpie Lark (*Grallina cyanoleuca*)
- Major Mitchell Cockatoo (*Lophochroa leadbeateri*)

- Masked Woodswallow (*Artamus personatus*)
- Mistletoebird (*Dicaeum hirundinaceum*)
- Mulga Parrot (*Psephotus varius*)
- Nankeen Night Heron (*Nycticorax caledonicus*)
- Owlet Nightjar (*Aegotheles cristatus*)
- Pacific Black Duck (*Anas superciliosa*)
- Painted Finch / Ilitja (*Emblema pictum*)
- Peaceful Dove (*Geopelia striata*)
- Pied Butcherbird (*Cracticus nigrogularis*)
- Pied Honeyeater (*Certhionyx variegatus*)
- Rainbow Bee-eater (*Merops ornatus*)
- Rufous Songlark (*Cincloramphus mathewsi*)
- Red-Backed Kingfisher / Luunpa (*Todiramphus pyrrhopygius*)
- Sacred Kingfisher / Luunpa (*Todiramphus sanctus*)
- Singing Honeyeater (*Gavicalis virescens*)
- Southern Boobook / Morepork (*Ninox novaezeelandiae*)
- Spinifex Pigeon / Ipuru (*Geophaps plumifera*)
- Spiny-cheeked Honeyeater (*Acanthagenys rufogularis*)
- Splendid Fairy Wren (*Malurus splendens*)
- Spotted Harrier (*Circus assimilis*)
- Tawny Frogmouth / Piwi (*Podargus strigoides*)
- Torresian Crow / Kaanka (*Corvus orru*)

- Wedge-Tailed Eagle / Walawuru (*Aquila audax*)
- Western Bowerbird / Ikarka (*Ptilonorhynchus guttatus*)
- Whistling Kite (*Haliastur sphenurus*)
- White-Faced Heron (*Egretta novaehollandiae*)
- White-Necked Heron (*Ardea pacifica*)
- White-Fronted Honeyeater (*Purnella albifrons*)
- White-Plumed Honeyeater (*Ptilotula penicillata*)
- Willie Wagtail / Tjindri tjindri (*Rhipidura leucophrys*)
- Yellow-Throated Miner / Piil-piilpa (*Manorina flavigula*)
- Zebra Finch / Nyii-nyii (*Taeniopygia guttata*)

Mammals

- Dingo (*Canis familiaris dingo*)
- Euro / Hill Kangaroo / Common Wallaroo / Kulpirrpa (*Osphranter robustus robustus*)
- Black-Footed Rock-Wallaby / Warru (*Petrogale lateralis centralis*)
- Fat-Tailed Pseudantechinus (*Pseudantechinus macdonnellensis*)
- Sandy Inland Mouse (*Pseudomys hermannsburgensis*)

Feral Species

- Cat / Miiyawu (*Felis catus*)
- Red Fox (*Vulpes vulpes*)

Reptiles

- Centralian Carpet Python/Tatjalypa (*Morelia bredli*)
- Black-Headed Monitor (*Varanus tristis*)
- Long-Nosed Dragon (*Gowidon longirostris*)
- Perentie/Ngintaka (*Varanus giganteus*)

Amphibians

- Centralian Tree Frog (*Litoria gilleni*)
- Desert Tree Frog (*Litoria rubella*)

Invertebrates

- Australian Emperor (dragonfly) (*Anax papuensis*)
- Blue Skimmer (dragonfly) (*Orthetrum caledonicum*)
- Rosy Skimmer (dragonfly) (*Orthetrum migratum*)
- Scarlet Percher (dragonfly) (*Diplacodes haematodes*)
- Blue Damselfly (*Austrolestes sp.*)
- Lesser Wanderer Butterfly (*Danaus chrysippus*)
- Lemon Migrant Butterfly (*Catsopilla pomona*)
- Meadow Argus (*Junonia villida*)

- Potter Wasp (*Abispa ephippium*)
- Giant Centipede (*Ethmostigmus rubripes*)
- Shield Shrimp (*Triops australiensis*)

Index by Species

Index by Common Name

Made in the USA
Monee, IL
05 November 2026

67953741R00076